Genders and Sexualities in History
Series Editors: **John H. Arnold, Joanna Bourke and Sean Brady**

Palgrave Macmillan's series, Genders and Sexualities in History, aims to accommodate and foster new approaches to historical research in the fields of genders and sexualities. The series promotes world-class scholarship that concentrates upon the interconnected themes of genders, sexualities, religions/religiosity, civil society, class formations, politics and war.

Historical studies of gender and sexuality have often been treated as disconnected fields, while in recent years historical analyses in these two areas have synthesised, creating new departures in historiography. By linking genders and sexualities with questions of religion, civil society, politics and the contexts of war and conflict, this series will reflect recent developments in scholarship, moving away from the previously dominant and narrow histories of science, scientific thought and legal processes. The result brings together scholarship from contemporary, modern, early modern, medieval, classical and non- Western history to provide a diachronic forum for scholarship that incorporates new approaches to genders and sexualities in history.

Sexual Forensics in Victorian and Edwardian England is the first detailed historical analysis of the forensic sciences concerned with sexual violence. Victoria Bates argues that ideas about crime, consent, and the female body can be better understood by examining what happens in courtrooms when cases of sexual assault are being tried. Crucially, these legal cases also provide a new way of seeing how people in Victorian and Edwardian England understood the shifting boundaries between childhood and adulthood. Medical witnesses in courtrooms increasingly wielded immense influence on the outcomes of trials as well as affecting popular beliefs and attitudes towards sexual crimes and the status of children. Drawing on a formidable range of primary sources, including detailed case studies of legal sessions in the courts of Middlesex and the South West, Bates reflects on sexual crime as a social problem. In common with all volumes in the 'Genders and Sexualities in History' series, *Sexual Forensics in Victorian and Edwardian England* is a sophisticated contribution to our understanding of the past. It is an absorbing read.

Titles include:

John H. Arnold and Sean Brady (*editors*)
WHAT IS MASCULINITY?
Historical Dynamics from Antiquity to the Contemporary World

Valeria Babini, Chiara Beccalossi and Lucy Riall (*editors*)
ITALIAN SEXUALITIES UNCOVERED, 1789–1914

Victoria Bates
SEXUAL FORENSICS IN VICTORIAN AND EDWARDIAN ENGLAND
Age, Crime and Consent in the Courts

Heike Bauer and Matthew Cook (*editors*)
QUEER 1950s

Cordelia Beattie and Kirsten A. Fenton (*editors*)
INTERSECTIONS OF GENDER, RELIGION AND ETHNICITY IN THE MIDDLE AGES

Chiara Beccalossi
FEMALE SEXUAL INVERSION
Same-Sex Desires in Italian and British Sexology, c. 1870–1920

Roberto Bizzocchi
A LADY'S MAN
The Cicisbei, Private Morals and National Identity in Italy

Raphaëlle Branche and Fabrice Virgili (*editors*)
RAPE IN WARTIME

Susan Broomhall
AUTHORITY, GENDER AND EMOTIONS IN LATE MEDIEVAL AND
EARLY MODERN ENGLAND

Matt Cook
QUEER DOMESTICITIES
Homosexuality and Home Life in Twentieth-Century London

Peter Cryle and Alison Moore
FRIGIDITY
An Intellectual History

Lucy Delap and Sue Morgan
MEN, MASCULINITIES AND RELIGIOUS CHANGE IN TWENTIETH CENTURY BRITAIN

Nancy Erber, William A. Peniston, Philip Healy and Frederick S. Roden (*editors*)
MARC-ANDRÉ RAFFALOVICH'S URANISM AND UNISEXUALITY
A Study of Different Manifestations of the Sexual Instinct

Jennifer V. Evans
LIFE AMONG THE RUINS
Cityscape and Sexuality in Cold War Berlin

Kate Fisher and Sarah Toulalan (*editors*)
BODIES, SEX AND DESIRE FROM THE RENAISSANCE TO THE PRESENT

Christopher E. Forth and Elinor Accampo (*editors*)
CONFRONTING MODERNITY IN FIN-DE-SIÈCLE FRANCE
Bodies, Minds and Gender

Rebecca Fraser
GENDER, RACE AND FAMILY IN NINETEENTH CENTURY AMERICA
From Northern Woman to Plantation Mistress

Alana Harris and Timothy Jones (*editors*)
LOVE AND ROMANCE IN BRITAIN, 1918–1970

Dagmar Herzog (*editor*)
BRUTALITY AND DESIRE
War and Sexuality in Europe's Twentieth Century

Josephine Hoegaerts
MASCULINITY AND NATIONHOOD, 1830–1910
Constructions of Identity and Citizenship in Belgium

Robert Hogg
MEN AND MANLINESS ON THE FRONTIER
Queensland and British Columbia in the Mid-Nineteenth Century

Julia Laite
COMMON PROSTITUTES AND ORDINARY CITIZENS
Commercial Sex in London, 1885–1960

Marjorie Levine-Clark
UNEMPLOYMENT, WELFARE, AND MASCULINE CITIZENSHIP
"So Much Honest Poverty" in Britain, 1870–1930

Andrea Mansker
SEX, HONOR AND CITIZENSHIP IN EARLY THIRD REPUBLIC FRANCE

Nancy McLoughlin
JEAN GERSON AND GENDER
Rhetoric and Politics in Fifteenth-Century France

Jeffrey Meek
QUEER VOICES IN POST-WAR SCOTLAND
Male Homosexuality, Religion and Society

Jessica Meyer
MEN OF WAR
Masculinity and the First World War in Britain

Meredith Nash
MAKING 'POSTMODERN' MOTHERS
Pregnant Embodiment, Baby Bumps and Body Image

Tim Reinke-Williams
WOMEN, WORK AND SOCIABILITY IN EARLY MODERN LONDON

Linsey Robb
MEN AT WORK
The Working Man in British Culture, 1939–1945

Yorick Smaal
SEX, SOLDIERS AND THE SOUTH PACIFIC, 1939–45
Queer Identities in Australia in the Second World War

Helen Smith
MASCULINITY, CLASS AND SAME-SEX DESIRE IN INDUSTRIAL ENGLAND, 1895–1957

Jennifer D. Thibodeaux (*editor*)
NEGOTIATING CLERICAL IDENTITIES
Priests, Monks and Masculinity in the Middle Ages

Kristin Fjelde Tjelle
MISSIONARY MASCULINITY, 1870–1930
The Norwegian Missionaries in South-East Africa

Hester Vaizey
SURVIVING HITLER'S WAR
Family Life in Germany, 1939–48

Clayton J. Whisnant
MALE HOMOSEXUALITY IN WEST GERMANY
Between Persecution and Freedom, 1945–69

Gillian Williamson
BRITISH MASCULINITY IN THE *GENTLEMAN'S MAGAZINE*, 1731–1815

Midori Yamaguchi
DAUGHTERS OF THE ANGLICAN CLERGY
Religion, Gender and Identity in Victorian England

Genders and Sexualities in History Series
Series Standing Order 978–0–230–55185–5 Hardback   978–0–230–55186–2
Paperback
(*outside North America only*)

You can receive future titles in this series as they are published by placing a standing order. Please contact your bookseller or, in case of difficulty, write to us at the address below with your name and address, the title of the series and the ISBN quoted above.

Customer Services Department, Macmillan Distribution Ltd, Houndmills, Basingstoke, Hampshire RG21 6XS, England

Sexual Forensics in Victorian and Edwardian England

Age, Crime and Consent in the Courts

Victoria Bates
University of Bristol, UK

© Victoria Bates 2016

All rights reserved. No reproduction, copy or transmission of this publication may be made without written permission.

No portion of this publication may be reproduced, copied or transmitted save with written permission or in accordance with the provisions of the Copyright, Designs and Patents Act 1988, or under the terms of any licence permitting limited copying issued by the Copyright Licensing Agency, Saffron House, 6–10 Kirby Street, London EC1N 8TS.

Any person who does any unauthorized act in relation to this publication may be liable to criminal prosecution and civil claims for damages.

The author has asserted her right to be identified as the author of this work in accordance with the Copyright, Designs and Patents Act 1988.

First published 2016 by
PALGRAVE MACMILLAN

Palgrave Macmillan in the UK is an imprint of Macmillan Publishers Limited, registered in England, company number 785998, of Houndmills, Basingstoke, Hampshire RG21 6XS.

Palgrave Macmillan in the US is a division of St Martin's Press LLC, 175 Fifth Avenue, New York, NY 10010.

Palgrave Macmillan is the global academic imprint of the above companies and has companies and representatives throughout the world.

Palgrave® and Macmillan® are registered trademarks in the United States, the United Kingdom, Europe and other countries.

ISBN 978–1–137–44170–6

This book is printed on paper suitable for recycling and made from fully managed and sustained forest sources. Logging, pulping and manufacturing processes are expected to conform to the environmental regulations of the country of origin.

A catalogue record for this book is available from the British Library.

A catalog record for this book is available from the Library of Congress.

Typeset by MPS Limited, Chennai, India.

Contents

List of Figures   viii

Acknowledgements   ix

Introduction: Sex, Sexuality and Sexual Forensics   1

1   Knowledge: The Foundations of Forensics   22

2   Injury: Signs and the Sexual Body   42

3   Innocence: Chastity and Character   76

4   Consent: Violence and the Vibrating Scabbard   105

5   Emotions: Medicine and the Mind   132

6   Offenders: Lust and Labels   158

Conclusions: Medicine, Morality and the Law   190

Selected Bibliography   197

Index   201

List of Figures

I.1	Pre-trial statements for cases of suspected sexual crime	5
I.2	Conviction rates in true bills without guilty pleas	6
2.1	Verdicts in cases with medical evidence of genital injury	60
5.1	Links between bodily signs and emotional states	139
6.1	Verdicts in cases with prisoners under the age of 18, before the 1885 Criminal Law Amendment Act	165
6.2	Verdicts in cases with prisoners under the age of 18, after the 1885 Criminal Law Amendment Act	166

Acknowledgements

I would like to thank the Arts and Humanities Research Council, for providing funding for this research [AH/H019553/1]. My research would not have been possible without the help of the staff at London Metropolitan Archives, Gloucestershire Archives, Somerset Heritage Centre and Devon Record Office. I also appreciate the permissions granted by Oxford University Press to reproduce a small amount of a recent *Social History of Medicine* article, the Crown Copyright license to reproduce court material and Devon Record Office, Gloucestershire Archives, London Metropolitan Archives, Somerset Heritage Centre and South West Heritage Trust to cite court records.[1]

Friends and colleagues too many to mention have inspired me in many ways over the last few years, but some must be given special thanks for their input into this project. I am extremely grateful to Dr Sarah Toulalan, who has always made the time to give me academic support, career guidance, detailed feedback and extensive advice on my work. My colleagues at the University of Bristol were also very generous with their time and feedback at a 'Dead Author' session for my work in progress. Further thanks for academic guidance go to Professor Kate Fisher, Professor Mark Jackson and Professor Joanna Bourke, all of whom provided helpful comments on earlier versions of this book. Dr Neil Pemberton has also helped me to work through my ideas during a number of conversations about this book, and has always inspired me to give more creative conference papers. Finally, this book has benefitted greatly from the advice of anonymous reviewers and assistance of the editorial team at Palgrave.

This book is dedicated to those who have supported me in innumerable other ways, always with love and patience: Fiona and Paul Bates, and Sam Goodman.

Note

1. The Crown Copyright License is available at http://www.nationalarchives.gov.uk/doc/open-government-licence/version/2/.

Introduction: Sex, Sexuality and Sexual Forensics

One summer's evening in 1894, a young girl named Rose Buckland went to St George Park in Bristol for the purposes of listening to a live band and socialising with her friends. This enjoyable day out turned rapidly into a nightmare for the girl when, she claimed, two youths began 'pulling her about' and lifting up her clothes.[1] Buckland complained to her mother and a case was brought before a magistrate in October of the same year, but was dismissed before trial. The dismissal of this case at first seems surprising. It had a number of direct witnesses, with testimony that supported Buckland's claims to distress and her efforts to escape from the accused boys. However, witnesses also raised questions about the girl's status as a victim. The law on sexual consent assumed Buckland – as a girl aged 12 – to be sexually immature, both in behaviour and body, but witness testimony indicated that she had been flirtatious with boys in the park. One friend of the prisoners stated that Buckland was only 'pretending to cry' and that she had previously encouraged him 'to go and lie on her'.[2] This testimony destabilised Buckland's status as an innocent victim at the hands of two older boys.

Medical testimony provided a scientific basis for these concerns. A local physician named William Brown examined the girl and found no signs of genital or bodily injury. However, he asserted, injury was unlikely because Buckland was not a virgin before the alleged assault. In the view of this medical witness, she also did not present the expected physical characteristics of an immature 12-year-old girl. Brown noted that Buckland was 'very fully developed' physically and was menstruating at the time of the alleged assault. He declared that 'I should have judged her age to have been more than twelve years'.[3] This testimony could have been interpreted in two ways by the court: firstly, to imply that Rose Buckland's mother, who failed to provide a birth certificate to confirm that Rose was under the age of sexual consent, was lying about her age; secondly, to highlight a gap between Buckland's legal immaturity and her physical and social maturity. Whatever the final reason for the dismissal of this case, its medical testimony draws attention to ambiguities and tensions in the law on sexual consent. Whether or not Buckland was actually 12 years of age, she was not childlike in physical appearance or sexual behaviour. Medical witnesses and the courts implicitly acknowledged that some girls matured at a faster rate than others, despite the clear-cut law on sexual consent.

The Buckland case indicates that sexual forensics is about much more than the history of crime. In Victorian and Edwardian England, local courts were important spaces for articulating, contesting and intensifying tensions about sexual maturity and consent.[4] Medical testimony, along with responses to it, provides insights into contemporary attitudes to age, class, gender and sexuality. Medical witnesses did not simply report on signs of bodily harm, but also gave meaning to these signs. In the Buckland case, as in many others, medical testimony about the female body had moral as well as legal significance; it raised questions about the complainant's general character and degree of development. As a knowledge practice, sexual forensics emphasised the variability of female bodies and emphasised that girls reached sexual maturity at a wide range of ages. In so doing it unsettled the clear boundaries between childhood and adulthood presented in law.

Victorian and Edwardian sexual forensics operates at the intersection between legal, social and medical history. Sexual forensics became increasingly important in the nineteenth century, as medico-legal practice in general gained more recognition.[5] Medical practitioners had long contributed evidence in cases of suspected crime, but their roles grew with the formal development of forensic medicine as a field of investigation – which came later in England than in Europe. New laws of evidence, drawing upon the early modern turn towards empiricism as a model for 'facts', also fuelled the development of forensic medicine in the nineteenth century.[6] While hearsay and opinion had theoretically been dismissed from the English courts over the course of the eighteenth century, in 1782 scientific witnesses were officially granted the right to give an opinion within their own science.[7] The 'expert' opinion of a man of science was then formally marked out as evidence, which could be the basis of a legal 'fact' as determined by the jury, but the opinion of a layman was not.

A 'man of science' was not separated from wider society. Although the law marked out forensic medicine as more objective than other kinds of evidence, in practice medical witnesses often used their rights as 'men of science' to address implicitly middle-class concerns around age, class and gender. As Jill L. Matus notes, the historic 'relationship between scientific knowledge and cultural imperatives is one of interplay and exchange' rather than a story of the 'authority of science'.[8] In Victorian and Edwardian Britain, sexual forensics and the courts operated in the service of middle-class social interests not national, legal or political interests. Medical witnesses in court played a role in the construction and propagation of so-called 'rape myths' by interpreting the physical, and to a lesser degree the mental, characteristics of victimhood. Medical witnesses contributed to social stereotypes by testifying on subjects as wide-ranging as a complainant's sexual history, character, physical maturity and alcohol consumption.

This book seeks to show the value of studying sexual forensics in depth, rather than only as a small part of general studies of sexual crime. It draws evidence from cases of non-consensual sex tried as misdemeanours at the Middlesex Sessions and the Gloucestershire, Somerset and Devon Quarter Sessions between 1850 and 1914.[9] Taking the two regions of Middlesex and the South West avoids any temptation to treat London as representative of national trends. Middlesex fell under the jurisdiction of the Metropolitan Police who were unique in their extensive use of police surgeons during the period under study. Middlesex was also largely although not entirely urban, especially before 1889 when it included large areas of the East End of London.[10] The comparably, although also not entirely, rural nature of the south-west Quarter Sessions is indicated by the fact that the counties' key cities such as Exeter, Bristol and Bath were excluded; the county sessions were drawn only from outlying regions.[11]

That this book actually highlights few differences between the courts of Middlesex and the South West is not due to any fundamental resemblance between the locations. Instead, it is the result of similarities across time and place in courtroom 'scripts', which reflected and reinforced social stereotypes around the characteristics of 'real' sexual crime. The repetitious nature of courtroom testimony across Middlesex, Somerset, Gloucestershire and Devon is perhaps surprising, but historically significant. It can be explained by a range of factors including the selective reporting of national newspapers, long-term legal tropes, shared responses to shifts in the national law on sexual consent, and broad social concerns about the uncontrolled sexualities of young girls and boys. The geographically widespread nature of middle-class anxieties has implications for wider histories of Victorian sex and sexuality, prompting a reconsideration

4 *Sexual Forensics in Victorian and Edwardian England*

of the commonplace assumption that anxieties about sexual immorality focused on the literal and metaphoric 'dirt' of the city.[12]

Medicine played a key role in constructing and reinforcing courtroom 'scripts'. Medical practitioners brought international research, knowledge of local communities and the concerns of middle-class society to bear on their testimony. Unlike other offences, such as murder, medical testimony in cases of suspected sexual crime drew upon intimate examinations of living complainants and conversations with victims. The nature of this contact with victims, combined with a general tendency for medical men to reinforce middle-class values, meant that medical testimony sheds light on society as well as on science. Despite the value of court records in general, and medical testimony in particular, few scholars have considered sexual crimes in the courts or sexual forensics due in part to their omission from published Old Bailey proceedings. The few historians who have focused on sexual forensics in England have drawn their evidence primarily from medico-legal textbooks.[13]

The theory of sexual forensics must be supplemented with a study of court cases, the best surviving evidence of which lies in pre-trial statements: of 1424 surviving pre-trial statements from Middlesex and 789 from Gloucestershire, Somerset and Devon, 429 and 179 respectively included medical testimony. These pre-trial statements were taken at the magistrates' court or 'police court', the first point of contact after pursuing a prosecution, which was chaired by a magistrate but had no jury. At this point cases could be dismissed, tried summarily for certain charges (such as common assault) or committed to a grand jury to be approved for trial by judge and petty jury. When a magistrate passed a case forward to the Middlesex Sessions or Quarter Sessions for trial, pre-trial statements – including medical testimony – were also given to the grand jury and trial judge. These statements therefore provide a good sense of the testimony that witnesses would have repeated at trial, as witnesses could not deviate extensively from this recorded 'script' given at the magistrates' court. They do not include the prisoners' defence, questions posed to witnesses, or any cases dismissed before trial, but still allow unparalleled insights into Victorian and Edwardian sexual crime. They provide a rare opportunity to examine sexual forensics in practice, rather than only in theory.

Figure I.1 breaks down the distribution of court cases across the regions and years under study. Overall, there were 1700 complainants in Middlesex and 928 complainants in Gloucestershire, Somerset and Devon who pursued charges of attempted rape; attempted carnal knowledge of girls under the felony clause of sexual consent legislation, which was raised from ten to 12 during the period; carnal knowledge or

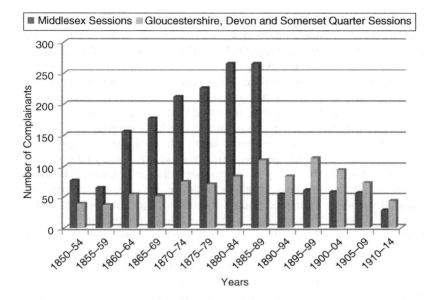

Figure I.1 Pre-trial statements for cases of suspected sexual crime
Note: (Attemped) rape, carnal knowledge, indecent assault, assault with intent and sodomy of a male under the age of fourteen

attempted carnal knowledge of girls under the misdemeanour clause of sexual consent legislation, which was raised from 12 to 16 during the period; indecent assault of males or females; assault with intent to commit a felony; and attempted sodomy of males under the age of 14.[14] The number of cases increased significantly over the course of the late-nineteenth century, dropping slightly again in the early-twentieth century. The only major decline in cases brought to trial in Middlesex after 1890 resulted from its reduction in size after the 1888 Local Government Act, rather than decreasing support for such cases.[15] The increase in trials over the period demonstrates the significance of late-Victorian and Edwardian England for histories of sexual consent and crime. These statistical trends tie in with national rates in trials for sexual crime, which rose from 1.5 per 10,000 in 1850 to 2 in 1880, reached a peak of 4 in 1885 and declined again to 2.5 by 1900.[16] Without entering into complex debates about the 'dark figure' of crime, such statistical shifts do not necessarily represent a 'real' rise in the extent of crime but rather higher rates of reporting and prosecution.[17]

The increase in trials for sexual offences was largely the result of late-nineteenth century campaigns to protect children, which raised the profile of sexual crime as a social problem and fuelled changes in the law on consent. Despite the rise in sexual offence cases, however, the conviction rate dropped dramatically in the same period. The question of why the conviction rate fell, at the same time as concern about sexual assaults increased, is central to this study. Sexual forensics operated against a backdrop of growing mistrust of complainants and anxiety about the rising age of consent, meaning that medical witnesses found success in the courts when they blurred the lines between adulthood and childhood; scientific evidence about physical maturity provided a counter-balance to the problematically clear law on sexual consent. Although the law showed growing sympathy for complainants in cases of sexual crime, increasing the age of consent to 16, the courts and medical witnesses expressed doubts about the innocence of girls under this age. They reinforced rape myths in which pubescent and adult females were untrustworthy unless absolutely chaste and flawless of character; against a flurry of social and legal change, courtroom 'scripts' and stereotypes actually became more embedded.

A downward trend in conviction rates in cases of suspected sexual crime, as depicted in Figure I.2, indicated shared medical and social

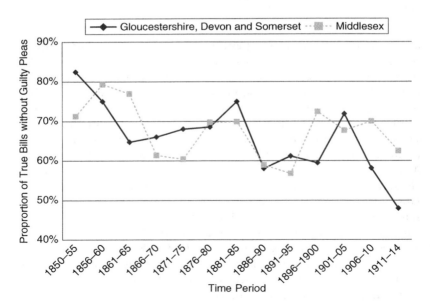

Figure I.2 Conviction rates in true bills without guilty pleas

concerns about age, character and consent. Similar patterns for Middlesex and the South West suggest that this decline in conviction rates was linked to broad trends in contemporary thought rather than being localised.[18] The dip in conviction rates in the 1890s corresponds to the findings of Louise Jackson at the Old Bailey, which she attributes to judges and juries 'running out of sympathy for girl victims' in the 1890s.[19] These trends were not only a post-1880s backlash, but also part of a long-term (albeit uneven) downward trend in conviction rates linked to earlier changes in sexual consent law and the growing social visibility of false claims and female precocity. This visibility came hand in hand with a child protection movement that, in seeking to increase the age of sexual consent for girls, created high-profile forums for discussion about issues such as precocity, blackmail and working-class immorality. Medical research on sexual maturity – which highlighted its lengthy and variable nature – had new implications in the light of growing concerns about working-class morality, respectability and the dangers of blackmail under new laws.

Crime statistics alone indicate that the period between 1850 and 1914 was crucial for concerns about sexual offences, public debates about the age of consent and growing middle-class concerns about working-class sexualities. As Stephen Garton notes, historians 'have pinpointed a crucial shift in Victorian culture around the middle of the [nineteenth] century' in terms of a growing middle-class emphasis on moral restraint and declining birth rates.[20] At the other end of this time frame the First World War marked declining interest in child protection and 'child abuse', also demonstrated by a drop in the number of cases reaching trial in the pre-war years, before the resurgence of such issues in the late-twentieth century.[21] While it would be a mistake to overstate the uniqueness or homogeneity of any given period, the years between 1850 and 1914 are notable for a growing interest in age, gender and sexual morality. This growing interest in sexual morality fuelled concerns about sexual crime, both in general culture and within the medical profession. Most such concerns, both in and out of the courts, focused on female victimhood and the female body: 87 per cent of pre-trial statements from Middlesex and 97 per cent from Gloucestershire, Somerset and Devon involved female complainants.

Overlapping concerns about child protection, crime prevention, gender roles and sexuality became central to society, law and medicine in the late-nineteenth and early-twentieth centuries. Due to these intersections, histories of medicine and sexual crime in Victorian and Edwardian England have much to contribute to the historiographies of gender and sexuality.[22] Sexual offences – along with the social, legal and medical responses to such crimes – are not transhistorical phenomena; sexual crime has always been socially constituted and interwoven with moral concerns about sexual behaviour. Medical testimony, with its attention to the bodies of victims and perpetrators, shaped and reinforced wider contemporary thought on gender and sexuality. Due to the dominant social profiles of doctors (middle-class males) and complainants (working-class females), medical testimony sheds particular light on middle-class male ideas about working-class femininity. Medical testimony, which focused on aberrations from the physical and mental norm, provides insights into Victorian and Edwardian ideas about 'normal' womanhood at different life stages; 'abnormal' signs only made sense in relation to the 'normal' body or behaviour. For contemporaries these norms represented not only the typical female but also the *ideal* female, in both body and behaviour; physical 'womanhood' was thus inextricably woven with embodied and performed 'femininity'.[23]

The centrality of bodily and developmental norms to sexual forensics could lend itself to a (late-) Foucauldian analysis, in which 'normalization' constitutes a form of discursive power.[24] This book touches upon such questions, as it shows that sexual forensics often provided a scientific validation for moral concerns about age, gender and sexuality. However, it also treats medical practitioners as a part of middle-class culture, without assuming that they had a distinct 'discourse' or any kind of discursive authority. Focusing on the authority of medical men over the female body, although an important part of understanding the history of sexual crime, can be a reductionist approach; in these frameworks, medical histories of sexuality often become histories of gender and power relations. This book seeks instead to consider how a range of factors, including ideas about male and female sexual maturity, informed the knowledge practices evident in Victorian and Edwardian courtrooms. It explores issues such as the value of ambiguity, as well as scientific objectivity, and the co-production of knowledge between medicine/law and doctor/patient, as well as the so-called medical 'gaze' on the female body.

Most existing histories of sexual offences focus on relationships between sexual crime, gender and social class; sexual forensics highlights the value of looking through the additional lens of age. There was no clear distinction between 'child' and 'adult' (or 'woman') in medicine, but instead a preoccupation with the transitional period between the two and an emphasis on the variable nature of sexual maturity. Sexual forensics shows that the prevalent historiographical separation of 'rape' from 'child sexual abuse', which represents current-day concerns more than historical ones, is misleading in its clear-cut divisions between sexual crimes against 'children' and against 'adults'. Scholars working on the history of childhood have increasingly distinguished between children of different ages, moving away from the concept of 'childhood' as a homogeneous or clearly delineated category, but few have yet applied such nuanced conceptions of age to histories of sexual crime.[25] This book provides an alternative medico-legal history of sexual crime that uses age as an analytical category, rather than making assumptions about the existence of a child/adult binary; it includes cases with complainants both above and below the legal age of sexual consent, making no clear distinction between 'childhood' and 'adulthood'. As Anna Davin notes, 'ultimately childhood can only be defined in relative terms. The question "What is a child?" must be followed by further questions: in whose eyes? When? Where? What are the implications?'[26]

By resisting a temptation to take the legal divisions between childhood and adulthood as representative of wider society, historians can better understand the complex frameworks of thought surrounding sex and sexual crime. With its focus on the body and life cycles, medical testimony reveals the particular anxieties that surrounded girls, and to some degree boys, at different ages. Medical witnesses focused particularly on complainants close to the age of puberty, including older children at the brink of this life stage. Girls at puberty constituted a high proportion of complainants, in part because puberty was a long process that occurred at a wide range of ages.[27] Sexual forensics was also particularly interested in puberty because of the ways in which the body, and therefore the meaning of bodily signs, changed with this life stage. Puberty represented an age of emerging gender difference and of sexuality, particularly in contrast with sexless infants and postmenopausal women. Medical testimony about puberty also connected to and consolidated wider middle-class concerns about working-class sexuality and gender, as puberty and 'girlhood' represented ages of potentially uncontrolled sexuality.

Bringing age into histories of sexual crime contributes to a burgeoning literature on childhood, puberty and adolescence. Scholars have studied extensively the Victorian and Edwardian fascination with 'girlhood' and adolescence, particularly in cultural histories, but puberty as a physiological process has often been neglected in such studies.[28] Physical aspects of adolescence have also received limited attention in scholarship. Adolescence is instead strongly associated with either middle-class education, as in the work of John R. Gillis, or with the psychologist G. Stanley Hall who propagated the term in 1904.[29] This general cultural, social and psychological focus on girlhood and adolescence overlooks medical models of sexual maturity. Such models, including some aspects of Hall's work, merit closer attention for the ways in which they interwove the social, physical and mental. These models also shifted over the late-nineteenth century with studies of child development and growing attention to developmental norms.[30] Sexual forensics provides important insights into the wider implications of such new forms of knowledge, beyond specific fields such as child development studies.

Medical interest in age and sexual development connected with wider social interest in these issues, which were part of high-profile conversations about child protection in the Victorian period. The English Parliament passed extensive child protection legislation over the course of the nineteenth century, most of which required decisions about when a 'child' became an 'adult'. Multiple pieces of legislation regulated child labour in industries ranging from mining to agriculture, increased the age of schooling and sexual consent, and provided for child protection from neglect.[31] In order to implement this legislation the state made widespread efforts to negotiate the differences between medical, legal, social and economic models of adulthood. However, there was no strict consistency in how childhood was defined in different legal contexts. The compulsory school-leaving age was established at ten in 1880 and raised to 11 in 1893, for example, by which time the age of sexual consent was 16 for girls. Childhood had to be redefined for each different law, a process that required negotiating the balance between society and biology in defining childhood. Debates about the age of transition to adulthood were ubiquitous and continuous in the late-Victorian years.

The Victorian flurry of legislative activity was the result, in part, of a shift in general attitudes to childhood. As part of the process of empire-building, over the course of the nineteenth century the state increasingly emphasised the value of social order and the importance of children to the nation's future. Such ideas were part of general middle-class consciousness, rather than merely political rhetoric, and soon also permeated children's culture from adventure tales to sports. Romantic ideals of childhood innocence also fed into campaigns for child protection in the nineteenth century. Fuelled by American theologian Horace Bushnell's internationally influential *Christian Nurture*, which was published in England in 1861, late nineteenth-century Protestantism increasingly emphasised that child development was a linear process, made up of developmental stages.[32] This religious context formed an important backdrop for medical research into child development and the age of puberty, as one such life stage. This framework – although not universally accepted – was also significant for its rejection of 'original sin', making puberty instead the age at which the onset of sin could occur. Such social and religious trends fed into the emergence of girlhood as a social category in the nineteenth century, which – like the nascent concept of adolescence – marked a period of instability before adulthood when personalities apparently became fixed. With 'original sin' now a contested category, children were deemed more than ever to need both protection and control.

Alongside growing social, legal and medical attention to the subject of childhood innocence, and its absence, in the late-nineteenth century sexual morality in general was under the microscope. Concerns about sexual offences were articulated primarily in relation to children and pubescent girls, but were part of a wider social shift in ideas about sexual behaviour and masculinity. Promiscuity, including the use of (young) prostitutes, came to represent a lapse of self-control at the same time that self-control became a marker of civilisation. The valorisation of self-control was not only the result of changing moral codes, but was also linked to economic ideals and to the individual focus – from self-help to self-discipline – of Victorian liberalism. Self-control of sexual behaviour was situated in opposition to the animal instincts of so-called 'lower' races, a belief that connected to literature on the British Empire and to new scientific theories of evolution and heredity. These ideas had implications for contemporary thought on uncontrolled masculine sexualities, which came to be associated less with rakishness and more with working-class immorality. Urbanisation and industrialisation further fuelled general middle-class anxieties about the lapse of self-control amongst working-class populations. Concerns about the impact of the city on morality fuelled a new connection between poverty and 'dirt' or moral contamination, particularly focused around the corrupting influence of urban environments on the innocent and the young. Such language was not entirely new, echoing early modern use of pollution as a metaphor to describe rape.[33] However, it took on a new significance in Victorian and Edwardian society in which growing cities came to represent chaos and immorality.

New frameworks for thinking about sexual morality and childhood fed into discussions about the age of sexual consent, among both those who advocated and those who feared change. These campaigns differed from anti-rape and 'child sexual abuse' campaigning in the late twentieth-century women's movement. The umbrella term of 'child sexual abuse' was never used in the nineteenth century and is not equivalent to the diverse historical terms used to describe sexual offences against children. As sociologist Carol Smart rightly notes, in relation to intergenerational sex in the mid-twentieth century, 'we should not assume that this variety of terms simply reflected different ways of saying the same "thing"'.[34] It was only in the late-twentieth century that 'paedophilia' and 'child sexual abuse' became a social focus. These concepts have emerged from campaigns to combat 'stranger danger' myths and to highlight the dangers that face children within the home. The moral panic of the late-twentieth century thus differed somewhat from that of the late-nineteenth century, which focused on defining and constructing the 'victim' rather than the perpetrator or the crime. For the Victorians, the age of sexual consent had implications for a wide range of issues seen as crucial to society including childhood innocence, pubescent instability, the future of the country and the need for male self-control. Although the interest in sexual consent came from a range of different voices, nearly all were driven by an interest in preservation of social order, morality and the family. Social purity campaigners viewed legislation as a means to enforce control over girls' morality at the 'dangerous years', while early women's movements focused on child protection.[35] Medical practitioners also had a voice in debates about sexual consent, often referring to research on ages of sexual maturity and development. Despite differences between these groups, they worked together because their respective positions emerged from a shared – and widespread – social concern about the importance of maintaining innocence in the young.

The most famous Victorian voice calling for change was that of W. T. Stead, editor of the London-based *Pall Mall Gazette*. Stead was the first to utilise the prevalent, but often implicit, concern about child welfare to create a widespread moral panic around sexual crime. On 6 July 1885, Stead launched an extensive and shocking exposé of child prostitution and 'white slave' traffic in his newspaper. Publishing under the title of 'The Maiden Tribute of Modern Babylon', Stead used ancient Babylon as a metaphor for London, describing how innocent female virgins were sacrificed to corrupt male 'Minotaurs'.[36] His ultimate aim was to raise the age of consent from 13 to 16 by forcing the passage of a languishing Criminal Law Amendment Bill through Parliament. The *Pall Mall Gazette's* sensationalist style of 'new journalism' described the seduction of young girls in unprecedented and melodramatic detail.[37] Most famously, Stead bought a 13-year-old girl for the purposes of proving that it was possible and was later imprisoned for assault in consequence.

The *Pall Mall Gazette* articles created uproar. Campaigner Josephine Butler noted in her private correspondence that 'I never saw anything like the excitement in the streets and in the House of Commons'.[38]

Medicine and the female body were an important part of Stead's exposé. Not only was a medical practitioner central to the scandal surrounding the 'Maiden Tribute', as the gynaecologist Dr Heywood Smith put the abducted girl under chloroform to certify her virginity, but Stead also consulted and cited medical opinion about sexual consent in the 'Maiden Tribute' articles. He quoted surgeon Frederick Lowndes, who would also write on the age of consent in *The Lancet*, as saying that 'many members of the medical profession, including myself, would wish to see an extension of the age [of consent] in females … carnal knowledge of any female under puberty is a cruel outrage'.[39] The medical dimensions of the 'moral panic' thus focused on questions around age, maturity and the body rather than on prostitution and the trade in girls. These issues had long been debated in the courtroom, but were now brought into public view.

Stead's exposé fanned the flames of concerns that had been growing for decades. The age of sexual consent was raised soon after, from 13 to 16, in the 1885 Criminal Law Amendment Act. This law also dealt with prostitution and sexual relations between men, but the age of sexual consent was a particularly dramatic and controversial change; it would have significant implications for cases of sexual crime in the courts and for sexual forensics. The 1885 law was the most high profile of five pieces of sexual consent and incest legislation passed between 1861 and 1908. These legal changes built upon Roman law, canon law, the 1275 Statute of Westminster and a further statute of 1576 that distinguished child victims; together these laws linked maturity to puberty and created a tiered system of sexual consent.[40] The 1861 Offences Against the Person Act consolidated the common law on sexual offences, which classified the carnal knowledge of girls as a felony if they were under the age of ten and as a misdemeanour if they were under the age of 12. In 1875 a new Offences Against the Person Act raised the felony and misdemeanour clauses of sexual consent legislation to 12 and 13 respectively. The law on indecent assault was also altered in 1880, in a Criminal Law Amendment Act that removed consent as a defence for prisoners who committed indecent assaults against boys or girls under the age of 13. In 1885, the better-known Criminal Law Amendment Act raised the aforementioned felony clause for carnal knowledge to 13 and the misdemeanour clause to 16. The 1908 Punishment of Incest Act also had some age-based elements and included lineal relationships such as father-daughter or uncle-niece incest, although the law was primarily passed to criminalise incest between consenting adults.[41] Most of these laws focused on girls due to particular concerns about the moral implications of a woman 'falling'. Boys were, however, protected under 'indecent assault' law and were implicitly protected up to the age of 14 in sodomy cases as the law considered them to be impotent, 'passive' participants in sexual acts before this age.[42]

There were many changes to the law on sexual consent over the course of the late-nineteenth century, but girls in the 'misdemeanour' category proved consistently to be the most problematic age group. Despite legal efforts to delineate ages of sexual maturity, the two-tier legal system (felony/misdemeanour) created a category of victims who were neither adult nor child. There was even a clause in the 1885 Criminal Law Amendment Act that the 'reasonable belief' that a girl was over the age of 16 constituted a defence. This clause was a concession to those who feared that girls would use the law for purposes of blackmail and extortion. Despite the law's claims to clarity, the framework implicitly allowed sexual capacity to be evaluated on a case-by-case basis rather than assumed on the grounds of age alone. This approach would form the basis of most medical witness testimony in court, which further blurred the grey areas between childhood and adulthood.

Although it had limitations and compromises, the highest profile piece of legislation – the 1885 Criminal Law Amendment Act – inspired legal changes around the globe. France, along with a number of countries that had adopted the Napoleonic code such as Portugal and Denmark, increased the age of consent from between ten and 12 to between 13 and 16 in the late-nineteenth century.[43] Many of these European changes were linked to a shared legal system, but social purity and early feminism were also of particular importance in shaping the law in the UK and USA. As Lynn Sacco notes, in her work on incest cases in North America, feminist groups conducted 'aggressive' campaigns for new age-of-consent laws and by 1900 32 of 45 states had set 16 as the age of sexual consent.[44] Age of consent legislation was also exported to a number of British colonies. India is perhaps the most famous example, where racial concerns and local religious tradition shaped modern rape law and debates around the Age of Consent Act of 1891.[45] In 1892 the *Women's Herald* also reported that the English legislation had an impact in Australia: '[a]fter years of agitation at last the Victorian Legislature hass [sic] passed a Criminal Law Amendment Act on the lines of that which has been in force in Great Britain for some years. It raises the age of consent to sixteen years'.[46]

Other countries, instead, responded to problems with existing sexual consent law by abandoning it. In Bolshevik Russia, for example, the concept of a strict 'age of consent' was replaced with the more flexible concept of 'sexual maturity'. Such actions sought to recognise the 'climatic ... racial ... sociocultural and individual conditions that influence the development of the child', a concern also evident in Victorian and Edwardian Britain, but Dan Healey has shown that the variable nature of sexual maturity meant that this 'elegant solution in theory ... proved extremely difficult to implement in practice'.[47] Healey finds that medical practitioners were actually ambivalent about this policy, rightly believing that applying a 'medicalized threshold' to law would create confusion in practice.[48] Although Bolshevik Russia differed from Victorian England in innumerable ways, Healey's findings have some implications for understanding sexual consent in the latter context. Medical practitioners in nineteenth- and early-twentieth century England also never actively resisted the legal age of consent as a concept. They recognised that medicine and the law operated in different ways and never sought to conflate the two frameworks; indeed, many medical practitioners actively advocated raising the age of consent in 1885. However, they appreciated that the courts allowed space for some ambiguities and moral concerns about girls' sexuality.

The following chapters focus on these ambiguities and moral concerns, particularly in relation to medical theories about sexual maturity. The first chapter outlines shifts in the nature of medical knowledge about developmental norms in the nineteenth century, before the book considers how such knowledge fed into sexual forensics in the form of medical evidence on bodily injury, chastity, resistance, emotional states and the perpetrator of sexual offences. Together, these chapters demonstrate that the judicial system encouraged and responded most positively to medical evidence when it reinforced moral stereotypes and rape myths.[49] In addition to general moral principles, grounded in religion, in the late-nineteenth and early-twentieth centuries rape myths drew upon shifting ideas about modernity, precocity, social order, gender roles and the future of the empire. Medical testimony that reinforced the 'normality' of the innocent child and demonstrated anxieties about the uncontrollable and unpredictable nature, and timing, of puberty supported these wider frameworks of middle-class thought. Moral and medical thought was neither homogeneous nor static in the Victorian and Edwardian years, quite the reverse. However, medical witnesses in court integrated contemporary cultural concerns about age, class, gender and precocity with new developmental physiology and traditional ideas about the sexual body, in a way that served to reinforce the status quo and to consolidate long-held stereotypes of chaste female victimhood.

Each of the final five chapters is followed by a single transcript and case analysis. This structure seeks to avoid a common pitfall of medico-legal histories in which, as noted in David Phillips' work on the history of rape, 'the more cases one deals with, the less opportunity there is to dissect in detail any individual cases'.[50] The case studies provide in-depth analysis of one issue raised in the paired chapter, but do not represent representative 'model' cases or claim to deal with all issues of significance. 'Case analysis' chapters put medical testimony into the context of a full trial transcript and show that it differed on a case-by-base basis, as a counter-balance to more traditional narratives of continuity and change over time. They allow for valuable insights into how broad medical trends operated in relation to specific – and varied – social, legal and moral concerns.

Due to its many dimensions, the significance of sexual forensics is confined neither to the courtroom nor to history. The legacy of Victorian moral panics about age, sexual maturity and sexual crime is still keenly felt. Most obviously, the age of sexual consent has remained at 16 for well over a century. New ideas about childhood and child protection in the nineteenth century have also taken root, as a Romantic notion of childhood innocence has come completely to replace Evangelical notions of original sin. This legacy is not entirely positive, however. Victorian concerns about the behaviour of precocious youths have also endured. In 2013, the UK's Crown Prosecution Service suspended a barrister for describing a 13-year-old victim of sexual crime as 'predatory'.[51] In 2015, a London judge handed down a suspended sentence to a prisoner because – she claimed – the 16-year-old girl 'groomed' her male teacher rather than *vice versa*.[52] The uproar around these cases indicate that such ideas are not as widely accepted as a century ago, but our society still carries echoes of Victorian and Edwardian thought.

The continued prevalence of victim blaming is also an unfortunate echo of Victorian anxieties around uncontrolled female sexuality. Over a hundred years after Rose Buckland was allegedly assaulted in a Bristol park, Bristol City Council launched a campaign entitled 'this is not an excuse'.[53] Posters were displayed around the city to dispel rape myths around marital rape, alcohol consumption, women's clothing and female complicity in encouraging rape. The need for such posters indicates that rape myths have maintained their foothold in the public imagination. However, the Victorian and Edwardian periods also sowed the seeds of resistance to such victim blaming and the rejection of a so-called sexual 'double standard' for men and women. Victim blaming has historically been reformulated – and resisted – in response to contemporary anxieties; medicine has not been an objective bystander in this process, but instead has often provided scientific validation for prevailing middle-class thought on age, gender, class and sexuality. Understanding the history of these ideas, both medical and social, can help us better to realise the assumptions that underpin our own rape myths.

Notes

1. Gloucester, Gloucestershire Archives (GA), Pre-Trial Statements, John Doggett and George Young not tried (no bill) at the Gloucestershire Quarter Sessions in October 1894 for indecent assault, Q/SD/2/1894.
2. Gloucester, GA, Pre-Trial Statements, Doggett and Young, Q/SD/2/1894.
3. Gloucester, GA, Pre-Trial Statements, Doggett and Young, Q/SD/2/1894.
4. The book uses 'Victorian and Edwardian' England to include the period up to World War I, including four years of George V's rule.
5. For general medico-legal histories and works on other forms of forensic medicine in the nineteenth century see Ian A. Burney, *Bodies of Evidence: Medicine and the Politics of the English Inquest, 1830–1926* (Baltimore, MD: Johns Hopkins University Press, 2000); Ian Burney and Neil Pemberton, 'Making Space for Criminalistics: Hans Gross and Fin-De-Siècle CSI', *Studies in History and Philosophy of Science* 44 (2013), 16–25; Michael Clark and Catherine Crawford (eds), *Legal Medicine in History* (Cambridge; New York: Cambridge University Press, 1994); M. Anne Crowther, 'Forensic Medicine and Medical Ethics in Nineteenth-Century Britain' in *The Codification of Medical Morality: Historical and Philosophical Studies of the Formalization of Western Medical Morality in the Eighteenth and Nineteenth Centuries*, ed. Robert Baker, vol. 2 (Dordrecht; London: Kluwer Academic Publishers, 1995), 173–90; Thomas Rogers Forbes, *Surgeons at the Bailey: English Forensic Medicine to 1878* (New Haven: Yale University Press, 1985); Katherine D. Watson, *Forensic Medicine in Western Society: A History* (London; New York: Routledge, 2010).
6. For a more detailed discussion of the development of rules of evidence and how they relate to expert testimony over time, see: Tony Ward, 'Law, Common Sense and the Authority of Science: Expert Witnesses and Criminal Insanity in England, Ca. 1840–1940', *Social & Legal Studies* 6 (1997), 343–62, pp. 345–47; Anthony Good, 'Cultural Evidence in Courts of Law' in *The Objects of Evidence: Anthropological Approaches to the Production of Knowledge*, ed. Matthew Engelke (London: Royal Anthropological Society, 2008), 44–57, p. 45; Wilson Wall, *Forensic Science in Court: The Role of the Expert Witness* (Chichester; Hoboken, NJ: Wiley-Blackwell, 2009), p. 58; and Carol Jones, *Expert Witnesses: Science, Medicine and the Practice of Law* (Oxford: Clarendon Press, 1994), pp. 102–14. Thanks go to Dr Noah Millstone for directing me towards some reading on earlier periods, including John H. Langbein, 'The

Historical Foundations of the Law of Evidence: A View from the Ryder Sources', *Columbia Law Review* 96 (1996), 1168–1202 and Barbara J. Shapiro, *A Culture of Fact: England 1550–1720* (Ithaca: Cornell University Press, 2000).
7. Folkes v Chadd. (1782) 3 Doug KB 157. This judgement consolidated rather than created the 'expert' witness practice; Jones, *Expert Witnesses*, p. 57.
8. Jill L. Matus, *Unstable Bodies: Victorian Representations of Sexuality and Maternity* (Manchester: Manchester University Press, 1995), p. 7.
9. Misdemeanours required a trial by jury, but often received lower penalties than felonies – for example, being cases of attempted rape rather than rape. Felonies were tried at Assizes, for which there are few surviving records of relevance.
10. Louise Jackson, *Child Sexual Abuse in Victorian England* (London: Routledge, 2000), p. 24.
11. These areas were not completely cut off from the expanding urbanisation of the 1800s, as contemporaries bemoaned a declining number of the agricultural classes in all three counties; 'On the Decrease of the Agricultural Population of England, A. D. 1851–61, by Mr Purdy', *The Athenaeum*, 5 September 1863, 311–12, p. 311. Many areas that fell within Gloucestershire were mining regions or sites of wool manufacture, for example, and therefore partially industrialised. However, these areas still contained large agricultural regions such as the Cotswolds in Gloucestershire and Somerset's Taunton vale'; 'Rural Economy of the British Isles: Western Counties', *New England Farmer* 8 (1856), 276–77, p. 276.
12. For example see Seth Koven, *Slumming: Sexual and Social Politics in Victorian London* (Princeton, NJ: Princeton University Press, 2006).
13. For example Ivan Crozier and Gethin Rees, 'Making a Space for Medical Expertise: Medical Knowledge of Sexual Assault and the Construction of Boundaries between Forensic Medicine and the Law in late Nineteenth-Century England', *Law, Culture and the Humanities* 8 (2012), 285–304. Some historians have studied court cases, but paid only limited attention to their forensic dimensions; for example, there is one chapter on medicine in Jackson, *Child Sexual Abuse*.
14. The statistics include those cases that were found a 'no bill' by a grand jury and those cases that do not have surviving pre-trial statements. The phrases 'sexual crime' and 'sexual offences' are used as umbrella terms to cover all of these types of non-consensual sex, but without claiming that they were synonymous. They all carried very different social, legal and medical meanings that this book considers where relevant. The crime of indecent exposure is excluded, as in the courts under study it had no relevance to medical practitioners. The 'exhibitionist' became a psychiatric category in the late-nineteenth century, but in Middlesex and the South West was not an issue upon which medical witnesses were consulted; Angus McLaren, *The Trials of Masculinity: Policing Sexual Boundaries, 1870–1930* (Chicago: University of Chicago Press, 1997), p. 194. On the late nineteenth-century development of 'exhibitionism' as a psychiatric category see also Joanna Bourke, *Rape: A History from 1860 to the Present Day* (London: Virago, 2007), pp. 250–51. The legal category of incest is similarly omitted from the list because after 1908 the crime generally related to sexual relations between consenting adults. Most cases of incest between adults and children were incorporated

in the category of 'carnal knowledge' of girls under the age of consent at the Quarter Sessions. Finally, these parameters exclude prostitution as a subject of close analysis. With the exception of procurement or brothel-keeping charges, it was not explicitly illegal and few prosecutions involved medical evidence.
15. 31,484 acres of the County of Middlesex were transferred to the County of London. These were mainly heavily urbanised regions, which included 2,697,271 people; British Parliamentary Papers (BPP), Minutes of Evidence: Royal Commission on London Government, Part 1, 1922, p. 377.
16. These figures incorporate all cases (tried on indictment) of: rape or unlawful carnal knowledge; attempted rape or unlawful carnal knowledge; assault with intent to rape; and indecent assault. They are taken from Figure 1.2 in Jackson, *Child Sexual Abuse*, p. 5.
17. For further reading on the 'dark figure' of (sexual) crime and the use of this term, see David Bentley, *English Criminal Justice in the Nineteenth Century* (London: Hambledon Press, 1998), p. 18; Bourke, *Rape*, pp. 15–18; Shani D'Cruze and Louise A. Jackson, *Women, Crime and Justice in England since 1660* (Basingstoke; New York: Palgrave Macmillan, 2009), pp. 28–29; V. A. C. Gatrell, 'The Decline of Theft and Violence in Victorian and Edwardian England' in *Crime and the Law: The Social History of Crime in Western Europe since 1500*, ed. V. A. C. Gatrell, Bruce Lenman and Geoffrey Parker (London: Europa, 1980), 238–70, pp. 286–89; Roy Porter, 'Rape – Does it have a Historical Meaning?' in *Rape: An Historical and Cultural Enquiry*, ed. Roy Porter and Sylvana Tomaselli (Oxford: Basil Blackwell, 1986), p. 221.
18. The maximum disparity between provincial and urban conviction rates was 15 per cent, in the period 1911–14, and in Figure I.2 the regions demonstrate similar peaks and troughs at key periods.
19. Jackson, *Child Sexual Abuse*, p. 106.
20. Stephen Garton, *Histories of Sexuality* (London: Equinox, 2004), p. 123.
21. Barry M. Coldrey, 'The Sexual Abuse of Children: The Historical Perspective', *Studies: An Irish Quarterly Review* 85 (1996), 370–80, p. 370.
22. See Garthine Walker, 'The History of Sexuality? A View from the History of Rape' <http://garthine.wordpress.com/2014/01/22/the-history-of-sexuality-a-view-from-the-history-of-rape/> (accessed 28 December 2014).
23. Definitions of 'womanhood' and 'femininity' remain contested, but within this book 'womanhood' is taken as a life stage and 'femininity' as a social role that also links to embodied experience and identity.
24. For a clear summary of Michel Foucault's theories on 'normalization' see James W. Bernauer and Michael Mahon, 'Michel Foucault's Ethical Imagination' in *The Cambridge Companion to Foucault*, ed. Gary Gutting (Cambridge: Cambridge University Press, 2005), 149–175, p. 151.
25. Works on childhood in which historians have taken a critical approach to age include Alysa Levene, 'Childhood and Adolescence' in *The Oxford Handbook of the History of Medicine*, ed. Mark Jackson (Oxford; New York: Oxford University Press, 2011), 321–37 and Sally Shuttleworth, 'Victorian Childhood', *Journal of Victorian Culture* 9 (2004), 107–13, p. 107. Stephen Robertson's work provides a rare example of nuanced approaches to age in relation to the history sexual crimes; Stephen

Robertson, *Crimes against Children: Sexual Violence and Legal Culture in New York City, 1880–1960* (Chapel Hill; London: University of North Carolina Press, 2005).
26. Anna Davin, 'What is a Child?' in *Childhood in Question: Children, Parents and the State*, ed. Anthony Fletcher and Stephen Hussey (Manchester: Manchester University Press, 1999), 15–36, p. 17.
27. Criminal trials for sexual offences generally involved complainants in their 20s at the oldest, with only four complainants in their 30s, one in their 50s, one in their 60s and two in their 70s.
28. For example Sally Mitchell, *The New Girl: Girls' Culture in England* (New York; Chichester; Columbia UP, 1995); Claudia Nelson and Lynne Vallone (eds), *The Girl's Own: Cultural Histories of the Anglo-American Girl, 1830–1915* (Athens; London: University of Georgia Press, 1994); Deborah Gorham, *The Victorian Girl and the Feminine Ideal* (London: Croom Helm, 1982).
29. John R. Gillis, *Youth and History: Tradition and Change in European Age Relations, 1770–Present* (New York: Academic Press, 1974); G. Stanley Hall, *Adolescence: Its Psychology and its Relations to Physiology, Anthropology, Sociology, Sex, Crime, Religion and Education*, vol. 1 (London; New York: Appleton, 1904).
30. On child development see André Turmel, *A Historical Sociology of Childhood* (Cambridge: Cambridge University Press, 2008), p. 84; James Wong, 'Critical Ontology and the Case of Child Development', *Canadian Journal of Political Science* 37 (2004), 863–82, pp. 873–76; Hugh Cunningham, *The Children of the Poor: Representations of Childhood since the Seventeenth Century* (Oxford; Cambridge, MA: Blackwell, 1991), pp. 192–97; Sally Shuttleworth, *The Mind of the Child: Child Development in Literature, Science and Medicine, 1840–1900* (Oxford: Oxford University Press, 2010), p. 9.
31. For detailed discussions of the various shifts in nineteenth-century child labour legislation and school-leaving ages see William Blackstone, *Commentaries on the Laws of England in Four Books with an Analysis of the Work* (Philadelphia: J. B. Lippincott & Co., 1867), pp. 348–49; and Eric Hopkins, *Childhood Transformed: Working-Class Children in Nineteenth-Century England* (Manchester: Manchester University Press, 1994).
32. For a discussion of Protestant ideas about 'linear human development' and the fixing of (religious) character after puberty, see Joseph F. Kett, 'Adolescence and Youth in Nineteenth-Century America', *The Journal of Interdisciplinary History* 2 (1971), 283–98; Joseph F. Kett, 'Reflections on the History of Adolescence in America', *The History of the Family* 8 (2003), 355–73. Bushnell's work was first published in the 1840s but later in England; see John Tosh, 'Authority and Nurture in Middle-Class Fatherhood: The Case of Early and Mid-Victorian England', *Gender & History* 8 (1996), 48–64, p. 53.
33. Garthine Walker, 'Rereading Rape and Sexual Violence in Early Modern England', *Gender & History* 10 (1998), 1–25, p. 2.
34. Carol Smart, 'A History of Ambivalence and Conflict in the Discursive Construction of the "Child Victim" of Sexual Abuse', *Social and Legal Studies* 8 (1999), 391–409, p. 393.
35. BPP, Report from the Selection Committee of the House of Lords on the Law Relating to the Protection of Young Girls, 1882, pp. 6–11. See also Ellice Hopkins's comments in 'The Criminal Law Amendment Bill', *The Standard*, 9 July 1883, 3, p. 3.

36. W. T. Stead, 'The Maiden Tribute of Modern Babylon – I', *Pall Mall Gazette*, 6 July 1885, 1–6, p. 1.
37. On melodrama see Judith R. Walkowitz, *City of Dreadful Delight: Narratives of Sexual Danger in Late Victorian London* (Chicago: University of Chicago Press, 1992).
38. London, The Women's Library, letter from J. Butler to Mrs Tanner and Miss Priestman, 10 July 1885, 3JBL.
39. W. T. Stead, 'The Maiden Tribute of Modern Babylon III: The Report of our Secret Commission', *Pall Mall Gazette*, 8 July 1885, 1–5, p. 3.
40. For further details see Julie Gammon, ' "A Denial of Innocence": Female Juvenile Victims of Rape and the English Legal System in the Eighteenth Century' in *Childhood in Question*, ed. Fletcher and Hussey, 74–95, p. 80.
41. Victor Bailey and Sheila Blackburn, 'The Punishment of Incest Act 1908: A Case Study of Law Creation', *Criminal Law Review* (1979), 708–18; Sybil Wolfram, 'Eugenics and the Punishment of Incest Act 1908', *Criminal Law Review* (1983), 308–16.
42. For more details of the law on sodomy and male impotence before the age of 14, see Chapter 6.
43. Stephen Robertson, 'Age of Consent Laws' in *Children and Youth in History: Item #230* <http://chnm.gmu.edu/cyh/teaching-modules/230> (accessed 28 February 2011).
44. Lynn Sacco, *Unspeakable: Father-Daughter Incest in American History* (Baltimore, MD: Johns Hopkins University Press, 2010), p. 50.
45. Elizabeth Kolsky, '"The Body Evidencing the Crime": Rape on Trial in Colonial India, 1860–1947', *Gender & History* 22 (2010), 109–30. For an example of the debates surrounding this legislation see 'Indian Child Marriage', *The Women's Penny Paper*, 26 July 1890, 474, p. 474.
46. 'Victoria', *Woman's Herald*, 27 February 1892, 7, p. 7.
47. Dan Healey, *Bolshevik Sexual Forensics: Diagnosing Disorder in the Clinic and Courtroom, 1917–1939* (DeKalb: Northern Illinois University Press, 2009), p. 10.
48. Healey, *Bolshevik Sexual Forensics*, p. 41.
49. Phrases such as 'rape myths' or 'real rape' have been used widely by scholars from a range of disciplinary backgrounds, generally in reference to stereotypes that 'real' rape involves violent assaults by strangers and active resistance on the part of a woman. Such 'myths' were long-held and did not generally originate in the courts, but were reinforced by judicial processes and by medical testimony. See Gethin Rees, ' "It is Not for Me to Say Whether Consent Was Given or Not": Forensic Medical Examiners' Construction of "Neutral Reports" in Rape Cases', *Social & Legal Studies* 19 (2010), 371–86, p. 372; Bourke, *Rape*, pp. 21–49.
50. David Philips, 'Sex, Race, Violence and the Criminal Law in Colonial Victoria: Anatomy of a Rape Case in 1888', *Labour History* 52 (1987), 30–49, p. 30.
51. Victoria Bates, ' "Predatory" 13-year-olds: We've Heard This One Before', <http://www.independent.co.uk/voices/comment/predatory-13yearolds-weve-heard-this-one-before-8751738.html> (accessed 29 December 2014).
52. 'Judge Criticised after Claiming 16-Year-Old Pupil Groomed Teacher', <http://www.theguardian.com/uk-news/2015/jan/14/former-london-teacher-convicted-over-affair-with-teenager>.
53. 'This is Not an Excuse' <http://www.thisisnotanexcuse.org.uk/> (accessed 29 December 2014).

1
Knowledge: The Foundations of Forensics

In May 1880, Harry Ernest Magrath was tried at London's Central Criminal Court at the Old Bailey for raping his seven-year-old daughter.[1] Upon being found guilty of the offence, the court sentenced Magrath to 20 years of penal servitude. A persuasive part of the case against Magrath came from three different medical witnesses, one of whom had a recognised specialism in forensic medicine and lectured on the subject. At the request of the Metropolitan Police, this forensic expert conducted a microscopical examination of the victim's linen and found 'blood cells' to corroborate the alleged offence. This case provides a representative example of high-level trials for sexual offences in Victorian and Edwardian London, in which it was common for a range of medical practitioners to testify and for those practitioners to have some access to laboratory facilities.[2] It also reflected the growth of medico-legal specialism, although specific training and employment in forensic medicine was largely limited to Edinburgh and London before the twentieth century.

The Old Bailey has been a focus of many historians' work on crime and forensic medicine, although few transcripts of sexual crime trials survive for the modern period.[3] This court is an important part of the history of forensics, as it often provided the site for extensive and specialist medical evidence. However, the Old Bailey differed greatly from the day-to-day practice of forensic medicine in mid-level courts. Mid-level courts tried indictable misdemeanours, such as attempted rape, instead of felonies, such as rape, or lesser crimes such as 'common assault' that came under the jurisdiction of police courts without a jury. The practice of these courts and the evidence given within them differed greatly from higher courts such as the Old Bailey. Despite the lower profile of mid-level trials such as the Middlesex Sessions, a high proportion of cases passed through these courts. The Old Bailey, which covered the City of London and the County of Middlesex, held 2476 trials for non-consensual sexual offences in the period 1850–1914 and the Middlesex Sessions alone held 1424 in the same period.

Courts such as the Old Bailey could draw upon cutting-edge diagnostic methods, but forensic evidence given in mid-level courts was more representative of the day-to-day practice of medicine in the Victorian and Edwardian periods. The specific value of these courts is indicated by a comparison between the above Old Bailey case and another tried at the Middlesex Sessions in the same year. The Middlesex court prosecuted an alleged indecent assault on a girl of the same age, seven years old, but differed in that only one medical practitioner testified in court.[4] He was consulted by the parents of the complainant, made no reference to use of a microscope and testified extensively on his discussion with the young girl and her understanding of the repercussions of lying (going 'to hell').[5] Such discussions of religion and non-scientific matters was by no means unusual in local and mid-level courts, particularly in the provinces where relationships between the police and medical practitioners were less formalised and there was only limited access to laboratory facilities. Such cases provide invaluable insights, but not into the scientific aspects of forensic medicine. Instead, they shed light on the intersections between medicine and general middle-class cultures and on shifting medical thought about age, gender and sexuality.

This chapter explores the knowledge base of sexual forensics in mid-level courts, with a particular focus on physiology – especially that related to sexual maturity – as the primary framework for diagnosing sexual offences. While higher courts shifted towards some bacteriological and chemical approaches to the body, as well as bringing in expertise on the mind over the late-nineteenth and early-twentieth centuries, most local courts' evidence base remained firmly grounded in the body. Rather than shifting towards the use of new specialisms, medical witnesses at the Middlesex Sessions and county Quarter Sessions drew upon new understandings of variable bodily norms and the age of sexual maturity in order better to understand the 'abnormal' body. They drew particularly upon the growth of anthropometric and social science research, which in the nineteenth century fed into a growing interest in the statistically typical population.[6] The 'normal' as typical, in addition to healthy and ideal, had a significant impact on understandings of the body in general and more specifically on medical evidence in cases of suspected sexual crime. In 1914 there was still a school of medical writing that focused on selected individuals, anecdotes or exceptional cases, but over the course of the nineteenth century statistical methods had increasingly supplemented – and in some contexts replaced – the traditional case report.[7] For sexual forensics, bodily norms created a new evidential framework. Although medical practitioners had long identified typical ages of puberty in broad terms, statistics provided a scientific basis for claims about relative rather than absolute forms of 'normality'; if a girl reached puberty later than the norm or was well developed 'for her age', it changed the meaning of her physical signs. Statistical norms allowed for a more nuanced approach to the body, taking into account its variability and recognising a law of errors rather than only averages.

Medical witnesses needed to draw upon an understanding of the 'normal' body at different ages and developmental stages in order to give meaning to physical signs. Beyond high-profile London courts, sexual forensics was generally not a pursuit of trained expert witnesses. Many lacked laboratory facilities as English police had no in-house forensic laboratories until the late 1930s and no specific 'rape examination suites' until the 1980s.[8] Instead of shifting its focus to the laboratory in the Victorian and Edwardian years, sexual forensics remained grounded in the naked-eye diagnosis of bodily signs. Naked-eye diagnosis relied not on extensive tests, but instead on knowledge about the 'normal' body, which was unstable and ambiguous. Physiology was therefore the cornerstone of sexual forensics, particularly research on puberty at which bodily norms apparently changed significantly. Female sexual maturity brought the capacity for sexual consent and significant physical changes, which altered the meaning of signs such as bloodstains, genital dilation (both of which became potentially 'natural' at puberty) and even marks of resistance as girls gained strength.

Books of medical jurisprudence, which guided medical testimony in court, indicate the significance of puberty for sexual forensics. Francis Ogston, a Scottish expert in forensic medicine who held a chair in medical jurisprudence, divided his 1878 book *Lectures on Medical Jurisprudence* into '1st, The violation of the female *under* the age of puberty. 2d, After puberty, and previously to her having otherwise had sexual connection. 3d, After puberty, and when the person had been accustomed to such connection'.[9] Ogston's work corresponded closely to that of Alfred Swaine Taylor, whose popular books of medical jurisprudence followed similar divisions throughout their many editions of the late-nineteenth century. Taylor divided his chapter on sexual offences into 'rape on children', 'rape on young females after puberty' and 'rape on the married'.[10] It was not until 1910 that Taylor's work reflected the law on sexual consent, by dividing into 'rape on infants and children up to the age of sixteen' and 'rape on women and girls over sixteen', rather than the physiological and social categories of puberty and marriage.[11]

Despite a general acceptance that the 'normal' (as typical) provided a useful framework for understanding the body, and by extension sexual crime, defining 'normal' puberty was not a clear-cut process. The lengthy and multifaceted nature of puberty posed problems for medical efforts to identify a single typical pattern of maturity. Physiology texts generally discussed sexual maturity in terms of three main features: physical maturity, such as the ability to reproduce; the ability to experience sexual sensations and pleasure; and sexual curiosity or the ability to understand sex.[12] A number of new theories emerged to explain the processes involved in sexual maturation over the course of the late-nineteenth and early-twentieth centuries, most significantly the concept of 'hormones' in 1905.[13] However, the central belief that puberty was a multifaceted process was unaltered. Across mainstream and specialist medical genres, authors conceptualised puberty as a 'gradual' process or 'evolution' that involved stages of physical and psychical maturation and that resulted in sexual differences being more clearly defined.[14] To complicate matters further, puberty marked the hastening of processes that could start many years earlier. There was no homogeneous category of 'childhood' within medical thought; the closer to the 'normal' age(s) of puberty, the less 'childlike' a girl or boy was thought to be.

In the nineteenth century, medical writers shifted away from early modern models of sexual maturity in which menarche (first menstruation) and sexual desire came together.[15] A specific order of development was put forward in many texts, which began with the emergence of secondary sexual characteristics including hair growth in both sexes, breasts in girls and vocal changes in males. Reproductive capacities apparently followed, prompting sexual sensations and the capacity for pleasure: for the male, physiologist William Carpenter wrote in 1855, 'sensations ... may either originate in the sexual organs themselves, or may be excited through the organs of special sense'; and, for women, according to physician Elizabeth Garrett Anderson in 1874, the 'functions of womanhood awaken instincts'.[16] In turn, these sensations would apparently stimulate curiosity and bring an ability to comprehend the nature of sex, bringing finally the psychological aspects of sexual maturity. Alienist Thomas Clouston even argued that 'for years after puberty boys and girls are still boys and girls in mind'.[17] Despite this apparently clear pattern of maturity, sexual maturation was not expected to be only a linear process. Most medical authors implied, and in some cases stated explicitly, that development of the three component parts of sexual maturity involved phases of both parallel and non-parallel development.[18] As puberty was seen as a life stage at which the body and mind developed at different times and rates, with mental capacity coming relatively late, there could be a few years in which the sexual body had reached maturity without a girl or boy yet having the capacity to understand or control it. This perceived pattern of development explains why puberty was such a concern for Victorian and Edwardian society, which increasingly valorised self-control, and why it was difficult for medical practitioners to identify any single 'normal' age for sexual maturity.

Although many aspects of sexual maturity had implications for sexual forensics, statistical studies of puberty were only possible for a few specific signs. Medical writers paid little attention to male bodies, in part because they struggled to pinpoint a clear and measurable marker of male sexual maturity. Secondary sexual indicators such as hair growth, they emphasised, emerged over a lengthy period of time and presented no clear means of measuring a typical age of sexual development.[19] The reproductive capacity in boys was also difficult to identify because young and physically immature males could produce semen without spermatozoa.[20] For girls the onset of menstruation, menarche, was a clearer indicator of maturity than the first male emission of semen. Blood had a symbolic value both as an indicator of womanhood and as a form of 'pollution' that needed controlling.[21] It was also an important sign for practitioners of sexual forensics who would need to use knowledge of the age of menarche to distinguish between 'natural' and 'unnatural' bloodstains. Statisticians generally ignored other aspects of puberty, such as mental capacity, and kept to general observations about the relatively belated nature of psychological maturity. Only sexologists made any attempt to analyse when girls achieved the capacity for sexual pleasure or the ability to understand sex, with studies that were limited in scope because of the difficulty of measuring these features of maturity.[22]

A number of high-profile writers turned to statistics to identify the average age of menarche in the late-nineteenth century. These studies often found averages of between 14 and 15 for menarche, which broadly corresponded to ages cited in early modern texts before the rise of statistical forms of 'normality'.[23] New statistical studies also strengthened long-held beliefs about the variable nature of puberty along social and moral lines. When the surgeon John Roberton systematically compared female puberty in 'Hindostan' [sic] and England in the early-nineteenth century, one of the earliest statisticians to do so, he found a higher age of puberty in the latter but no clear link between climate and age of menarche.[24] Roberton pointed instead to social factors as the cause of lower ages of menarche in warm climates, including looser moral standards, earlier marriage and premature sexual intercourse.

Although statistical studies did not revolutionise thinking on the average age of sexual maturity, they showed that deviation from the norm was more common and more extensive than previously believed. Sara Read and Sarah Toulalan have both shown that early modern medical literature, although often emphasising that a divergence of a year or two was not unhealthy, repeatedly cited 14 ('or twice seven years') as the age at which menarche generally occurred.[25] Research in the nineteenth century confirmed these averages, but also showed the extensive deviation from them. Roberton used his study of menarche in 450 women to declare in 1851 that puberty 'occurs in a more extended range of ages and is more equally distributed throughout that range, than authors have alleged'.[26] In 1885 anthropometrist Charles Roberts emphasised that medical practitioners were 'deluded by mere averages' as only 75 per cent of women had started menstruation by the age of 16.[27] Roberts applied the statistical 'law of error' to emphasise that only 20 per cent of females actually arrived at puberty at the average age.[28] The medical community was generally receptive to this research; medical writers discussed such studies explicitly within professional literature and fed their findings implicitly into advice literature aimed at the general public.[29] In 1888, for example, George Black (M.B.) wrote in *The Young Wife's Advice Book* that 'in our own country many cases are met with in which a girl has begun to menstruate when she was ten or twelve years old, and others in which the monthly discharge has been delayed till the twentieth year, or even longer'.[30]

Despite a growing interest in identifying 'normal' development patterns, new statistical studies of puberty showed that deviation from the norm was widespread. The statistical distribution of menarche was a continuum with a gradual transition between the average and extreme outliers. In visual form this statistical distribution would have formed a bell curve or 'Gaussian curve', a mathematical concept newly applied to humans under the influence of statistician Adolphe Quetelet (the mentor of Charles Roberts) and eugenicist Francis Galton.[31] Whether statistics were actually presented in tabular or graphical form, they formed a normal distribution rather than identifying *the* 'normal' person. In consequence, statistical studies of sexual maturity revealed no clear line between 'normal' and 'abnormal' ages of sexual development.

If puberty began at the age of 11 or 12 it was on the boundary of the 'normal' range, falling close to the expected age of puberty but not close enough to be typical. As Georges Canguilhem argues in his seminal text on the concept of 'normality', 'statistics offer no means for deciding whether a divergence is normal or abnormal'.[32] Statistical studies of menarche proved that medical ambiguity was a legitimate part of scientific knowledge, rather than the result of a lack of research or understanding of the female body.

As the somewhat hazy concept of 'normality' came to combine the typical, healthy and ideal body, anxieties grew alongside about certain forms of – equally ambiguous – 'abnormality' such as precocity.[33] Although some medical statisticians followed Galton's approach in their later work, in which the 'abnormal' could be superior to the 'average', they never articulated such ideas in relation to sexual maturity: 'abnormally' early (precocious) sexual development was a repeated focus of medical and social anxieties. Precocity also had implications for sexual forensics, as it destabilised a complainant's physical and social status as an innocent child. Medical practitioners in court presented some precocious girls as complicit in their 'fall', even if they were under the age of consent in law. Sexual precocity was a unique concern of the period, as the term had previously been used mainly in the context of plants or in relation to early intellectual development.[34] It became increasingly visible over the course of the nineteenth century, in connection with a range of broader anxieties about modernity, urban living and the importance of children to the future of the nation.

Claudia Nelson even argues, in her work on *Precocious Children & Childish Adults*, that the role of girls in the Victorian economy meant that 'the working-class girl was always a precocious child-woman'.[35] Nelson also shows that in Victorian literature the economically precocious girl was treated relatively sympathetically, compared with the precocious boy and 'the arrested child-woman', but notes the greater ambivalence surrounding female sexual precocity within the context of marriage (the 'child wife'). For medical writers and social commentators of the late-nineteenth and early-twentieth centuries, however, 'precocity' generally did not refer to the economic forms or to 'child wives', but instead to the premature development of unmarried girls. The forms that such premature development took were multiple but were universally negative. Precocity generally represented concerns about sexual development or behaviour, to quote a contemporary physician, 'years before the proper time'.[36] This issue of the 'proper time' positioned the precocious girl outside of developmental and social norms, and disrupted the 'natural' order of things.

Precocity represented a plethora of social anxieties about sex, age and gender. As physician R. L. Langdon-Brown stated, at the Seventy-Fifth Annual Meeting of the British Medical Association in 1907:

> Speaking generally, one may say that the word carries a bad meaning ... The term 'precocity' has been used with very different meanings, and often with no very definite meaning at all, but in a vague, loose, or popular way, and some attempt to understand these possible differences seems necessary if discussion of the subject is not to fall into confusion. By some it is defined as an earlier than average attainment of the ultimate growth of maturity; by others, again, as an unduly rapid development in relation to some assumed norm. Whichever of these be accepted, it is important to observe that we have in our mind some standard for comparison.[37]

In positioning the precocious girl against the 'average' and the 'norm' in this way, medical practitioners helped to define her *opposite*: the typical, healthy and ideal young girl.[38] In comparing a girl's body to the 'norm' or the 'average', medical practitioners drew simultaneously upon an emerging form of scientific knowledge and upon middle-class anxieties about working-class childhood. Precocity continued to hold negative connotations well into the twentieth century, both within and beyond physiology. In a 1914 work entitled *Sex*, botanist Patrick Geddes and Natural History specialist J. Arthur Thomson did little to contest the negative connotations of 'precocity' when they compared 'the salmon with its years of growth' with 'precocious maturity as in a rat'.[39]

Changing understandings of sexual maturity provided an important backdrop to sexual forensics in the nineteenth and twentieth centuries, albeit in a form that fuelled rather than resolved ambiguity. Forensic knowledge, physiology and new research about developmental norms were produced, learned and used in a range of spaces. It is impossible to grasp fully the knowledge base of sexual forensics without understanding the ways in which medical practitioners gained this knowledge and how they adapted it to the particular requirements of the courtroom. At the most basic level, medical knowledge about physiology and sexual development came from medical education. By the late-nineteenth century the pathways for medical education had been narrowed, partly in response to the requirements of the 1858 Medical Act for professional registration. Medical practitioners who worked in Devon, Somerset, Gloucestershire and Middlesex often had a shared knowledge base as they did not come from entirely distinct educational backgrounds.

Many provincial practitioners had originally worked elsewhere and most registered practitioners had trained at the main London or Scottish institutions. Although Britain was generally behind Europe in developing forensic medicine as a specialism, medical jurisprudence had long been taught in Scotland and was a growing part of medical training in London from the 1820s onwards.[40] Medico-legal matters were also part of examinations in pathology and surgery for a Fellowship of the Royal College of Surgeons. Beyond specific training in medical jurisprudence, physiology and sexual development were taught universally. For those medical practitioners who had not trained specifically in the subject of medical jurisprudence, this knowledge would be crucial to their medico-legal practice. However, medical education systems did not respond rapidly to new knowledge forms and many medical practitioners had trained years before being called upon to testify in court. Education provided some shared grounding in the principles of physiology and sometimes medical jurisprudence, but was not the primary mode by which new theories of bodily norms reached practitioners of sexual forensics.

As Anne Digby notes, medical students in the nineteenth century 'were clear that a formal medical education would only serve as a preliminary to a lifetime of further learning during practice'.[41] Once medical practitioners finished their training, they did not abandon the pursuit of new knowledge. They became, in the words of one Middlesex witness, 'medical reader[s]' instead of medical students.[42] Many general practitioners and specialists alike would have been acquainted with changing ideas about puberty and bodily norms because of their wide dissemination in medical literature. Both through direct and indirect citation, statistical studies of puberty reached the pages of many important general medical journals and medico-legal texts. *The Lancet* and *British Medical Journal* both disseminated new statistical studies of puberty, sometimes in specific relation to the subject of sexual crime and consent. Medical journals were of unparalleled importance in creating and reflecting knowledge at all levels of the medical profession. Wide geographical distribution of these journals was possible because of reforms in transport and the postal system.[43] *The Lancet* had circulation figures of several thousand just two years after it was launched in 1823 and the *British Medical Journal* had a circulation of 30,000 by the early-twentieth century.[44] Rural and urban medical practitioners alike not only read but also contributed case reports or opinion pieces to the two journals.

The nature of knowledge published in these medical journals differed from that taught in medical schools. Instead of disseminating established knowledge, medical journals provided spaces for the discussion of new and unproven ideas. Articles in medical journals were often part of a dialogue and in some cases were highly opinionated, particularly those in *The Lancet*. Medical journals were at the forefront of changing knowledge, even though they often acknowledged the limitations of new research. By publishing the new statistical studies of puberty, alongside relevant discussions about the age of sexual consent, these journals facilitated their reaching a wide and general medical audience. The *British Medical Journal* published correspondence from James Whitehead about the work of John Roberton, which observed minor discrepancies in their findings on the average age of puberty in England.[45] In 1885, Charles Roberts drew upon his lengthy experience researching the 'normal' body to write about 'The Physical Maturity of Women' in *The Lancet*.[46] This article was part of a discussion about the pending Criminal Law Amendment Act and paid extensive attention to the relationship between physical maturity and sexual capacity. Although statistics on menarche were fallible, such journal articles implied, the principle of measuring and observing developmental norms was worthy of attention. They did not shift entirely away from the 'exceptional' cases that had marked earlier studies of development, with widespread case studies of precocious puberty or pregnancy still prevalent in medical journals.[47] However, by publishing such cases alongside statistical accounts, these journals began to show the inherent problems of defining the 'exceptional' and further destabilised any notion of clear-cut life stages.

Books of medical jurisprudence also propagated new ideas about the 'normal' body, in part because the professionalisation of forensic medicine in the nineteenth century coincided with the rise of statistical methods. Alfred Swaine Taylor's influential *Manual of Medical Jurisprudence* was first published in 1844, around the same time that the work of early statistical researchers such as John Roberton was circulating and gaining influence. In consequence, from the outset, medical jurisprudence was grounded in new concepts of the 'normal' body at puberty. Like physiologists and medical statisticians, authors of forensic medicine texts placed great emphasis on menarche as an indicator of maturity. Taylor used 'puberty' as his guiding principle for interpreting evidence in cases of suspected sexual crime and between the 1850s and 1890s cited an increasingly wide range of statistical surveys of menarche, first by 'Dr. Rüttel' and later by Francis Hogg and 'Dr. Cohnstein'.[48] Taylor also grounded sexual forensics in the flexible concept of puberty, with all of its inherent ambiguities, rather than in the more clear-cut law on sexual consent. In the 1870s he added a new observation that, although statistical studies revealed an average age of 15 for menarche, '[t]he earliest and latest period in a large number of cases were respectively 9 and 23 years'.[49] Like medical journals, books of medical jurisprudence provided knowledge about sexual development in a particular form. As Taylor acknowledged, '[m]edico-legal knowledge does not consist so much in the acquisition of facts as in the power of arranging them, and in applying the conclusions to which they lead to the purposes of the law'.[50] For writers of forensic medicine texts, sexual maturity and bodily norms were of interest primarily because of their implications for signs of sexual crime; the facts were therefore 'arranged' as such. Medical jurisprudence textbooks paid particular attention to the implications of puberty for the meaning of signs such as bloodstains and genital dilation, because these shifted from markers of violence to potential indicators of natural maturity at puberty. They also wrote on issues such as the links between puberty and false claims, which reflected the 'purposes of the law' and reinforced rape myths.

The importance of medical jurisprudence texts in shaping medical testimony cannot be overstated. *The Lancet* noted that certain members of the profession, such as house surgeons of hospitals, were particularly likely to keep up to date with medico-legal literature because they found 'medico-legal work profitable' and might work as lecturers on the subject.[51] Many general practitioners also used the texts in their training, especially if they took courses in forensic medicine at university, and kept copies for practice. These books became continued points of reference during medical practice, even though practitioners did not necessarily purchase the latest editions. Taylor's *Manual of Medical Jurisprudence* reached 12 editions between 1844 and 1891 and his *Principles and Practice of Medical Jurisprudence* reached six editions between 1865 and 1910. The first of these texts sold nearly 16,000 copies between 1844 and 1861 alone.[52] Medical witnesses in a wide range of cases were quizzed on their familiarity with medical jurisprudence literature, and generally were able to show knowledge of key texts. In trials for a range of different crimes at the Old Bailey, including sexual offences, medical depositions were littered with statements such as 'I have read Dr. Taylor's *Medical Jurisprudence*' and 'I have read authors upon medical jurisprudence – I recognise Beck and Taylor as high authorities upon that subject'.[53] Sexual forensics was no exception to the widespread influence of these books. In pre-trial statements for the Middlesex Sessions in 1868 and the Gloucestershire Quarter Sessions in 1898, for example, surgeons stated that '[i]t is stated in Books of Medical Jurisprudence that ...' and referred to 'the indications given in the legal books as evidence of rape'.[54] Members of the judiciary also used forensic medicine texts. Medical journals repeatedly commented in their reviews that the books were in the 'interest of the practitioner and of the lawyer' and that '[l]awyers, as well as doctors, find the work of value to them'.[55] A range of professions therefore drew upon medical jurisprudence texts, which came to represent shared medico-legal pools of knowledge about the body and sexual crime.

Medical practitioners gained further knowledge about sexual development and bodily norms through experience. Statistics provided a scientific base for many medical practitioners' pre-existing knowledge of the wide range of bodily norms at different ages, as witnessed in their daily practice in the hospital or clinic. Medical witnesses also gained an understanding of 'normal' development patterns by comparing bodies visually, rather than only drawing upon textbook knowledge. In Devon in 1876, for example, a local physician and surgeon examined a 10-year-old complainant and testified that rape was not possible 'on a girl of that age judging from her developement [sic]. I examined her in contrast with a child of nearly similar age and found no appreciable difference'.[56] This medical witness did not assume that a rape was impossible on the basis of age alone, but rather on the basis of comparing her with another girl. This examination showed the complainant to be a 'normal' 10-year-old girl both in developmental terms, proving that she was not precocious or capable of sexual activity, and in health. Medical definitions of bodily norms did not come only from textbooks, but also from clinical experience and visual comparison.

General medical knowledge about bodily norms also came from interactions with lay knowledge, particularly in the therapeutic space of the clinic. The hidden nature of sexual offences meant that cases often first came to the attention of relatives of complainants, 19 per cent of whom consulted medical witnesses *before* a criminal charge was taken to the police. In such circumstances, which typically involved parents of young complainants seeking advice, medical practitioners responded to lay concerns about 'abnormal' bodies and shared their own ideas in order to guide decision-making about whether or not to pursue a charge. Nancy Theriot and Willemijn Ruberg have shown that knowledge about the unhealthy or violated body was often co-constructed through interactions between doctors, patients and families in nineteenth-century Anglo-America and the Netherlands respectively.[57] The same trend was evident in Victorian and Edwardian Britain. Mothers often brought cases to the attention of medical men, based on their own observations of physical 'abnormality' in their children. Medical witnesses relied on parental knowledge of certain developmental markers in order to interpret physical signs, such as whether a girl had reached menarche, and on health histories such as whether a child suffered from any natural diseases.[58] The 'normal' body in this context was that of a single child, about whom knowledge was co-produced between doctor and layperson, but such cases had a cumulative effect. Medical witnesses who came to understand the bodily norms of different individuals, through their interactions with parents of children and long-term patients, could then draw upon that experience when evaluating the bodies of strangers in future. Lay definitions of bodily norms fed into medical knowledge of when certain bodily changes moved from 'healthy' to 'unhealthy' or from 'natural' to 'unnatural'.

Although medical witnesses in court spoke little of the quack literature from which professionals sought increasingly to distance themselves, as the term 'quack' was associated with unqualified healers or the selling of false cures, this form of medical knowledge still had great significance for laypeople in the nineteenth century. As late as 1911 a professor of gynaecology, writing for the *British Medical Journal*, bemoaned 'the rush of inhabitants from the country to the towns' that made people turn to quack medicine because they were 'more neurotic, querulous, and introspective'.[59] This article also emphasised that 'women are specially prone to functional nervous disorders, so well known to the gynaecologist, which are the special prey of the quack' and noted the tendency for women to turn to quackery for euphemistically-titled 'female irregularities'.[60] Nervous disorders, female 'irregularity' and gynaecological matters also all came within the province of general practitioners and sexual forensics. Quackery may have therefore had an indirect influence on general practice through medical interactions with patients, many of whom drew upon oral traditions, quack literature and medical advice books in identifying and treating signs or symptoms.

After education, reading and experience in the clinic, medical witnesses finally gained knowledge from the courts themselves. Although few medical practitioners testified repeatedly, with the exception of some divisional surgeons of police, the courts produced a particular type of medico-legal knowledge. Historians have long noted the artificial nature of criminal trials as evidence, speaking of them as mediated 'scripts' or 'staged events'.[61] The trends and tropes of these scripts were remarkably consistent over time and place, generally operating to consolidate rather than to challenge stereotypes of victimhood and sexual crime. Medical testimony served a similar purpose. The courts determined which witnesses and cases reached trial, even before the adversarial process shaped the nature of their testimony. The judicial process as a whole encouraged new medical knowledge on sexual maturity and allowed its ambiguities, but only when it reinforced existing social stereotypes and rape myths. Judges defined what forms of medical knowledge were legally relevant and, particularly in cases of suspected sexual crime, often sought to connect scientific with social issues. The adversarial process elicited a particular form of medical testimony that was not only legally relevant but was morally significant. Pre-trial statements were generally constructed from short answers given in court, for example '[d]id you begin talking? Yes' might have been written up as 'I began talking'; some answers were longer, but few were spontaneous. [62] The lengthy prose of pre-trial statements therefore was not always a direct monologue, but was part of a carefully constructed 'script'.

In combination, these different methods of knowledge production had some effect on sexual forensics in the Victorian and Edwardian courts. In line with general conceptual shifts in thinking about the body, and specific knowledge gained from a combination of experience, reading and training, the word 'normal' became increasingly conspicuous in nineteenth-century medical testimony. In the seventeenth and eighteenth centuries, medical testimony at the Old Bailey focused on whether bodily signs had a 'natural cause' or were the result of the 'natural symptoms that are incident to Women' rather than their 'normality'.[63] Comments about the 'normal' body emerged only in the nineteenth century and were specific to medical testimony. Between 1850 and 1914, medical witnesses in Middlesex, Gloucestershire, Somerset and Devon referred to the 'normal' or 'abnormal' body in court with increasing frequency: not a single medical witness used the word 'normal' or 'abnormal' in the 1850s, but seven did so in the period 1860–1880, 13 in 1880–1900 and six in the period 1900–14 (for which fewer transcripts survive). Although a relatively small proportion of the hundreds of medical witnesses who testified during these years, these references indicate a broader shift in ways of knowing. Normality did not replace 'natural' and 'unnatural' as ways of understanding the body, as 'normal' and 'natural' were often used interchangeably to describe signs not caused by human interference. However, norms – in the sense of typical patterns of development – added a new dimension to sexual forensics over the course of the nineteenth century. 'Natural' signs made sense only in relation to 'unnatural' signs, caused by human interference, while 'normality' and its opposite 'abnormality' shed light on wider ideas about what a female body would – and perhaps more importantly *should* – look like at different ages.

The explicit use of the term 'normal' was only a small part of a general shift in the tone of medical testimony. Medical witnesses often interpreted bodily signs against more implicit developmental norms. There was a difference not only in language, but also in meaning, between two Middlesex cases from 1886 in which medical witnesses stated that 'the vagina is dilated more than is usual in a child of her age' and 'it is not usual to find blood from so young a child', and an Old Bailey case from 1723 in which the medical witnesses deposed that 'the Privy Parts were much dilacerated, and extended beyond their natural Dimensions'.[64] Despite referring to the same bodily sign, the dilation of the private parts, these two statements were grounded in distinct knowledge forms. Medical reference to what was 'usual' and the statement 'for her age' changed an absolute measurement to a relative one in relation not only to her individual body, but also to a developmental norm. Discussing a particular sign as (a)typical or (un)usual at a particular age embraced the concept of norms but also acknowledged their variability.

Overall, the concept of 'normal' as typical was crucial to sexual forensics in a broad sense, but statistics in themselves provided no clear or consistent way for doctors to approach the subject of sexual crime. They showed that sexual development varied along social lines and that any developmental norm, such as the average age of menarche, fell at the centre of a wide 'normal' range. This range had boundaries, but such findings made it difficult for medical witnesses to interpret bodily signs as 'normal' or 'abnormal' at any given age. Despite these limitations, the concept and language of normality entered the practice of sexual forensics. Medical witnesses and the courts actually gave value to ambiguity, of bodily norms and their opposites such as precocity, as a form of knowledge practice. Normality had value as both a scientific concept and a means to articulate middle-class moral concerns or bodily ideals. Physiological research also provided a scientific base for medical tendencies to assess sexual maturity on a case-by-case basis, rather than using a single template of age, bodily signifiers and what they signified. Reading the body was, in practice, as much moral as medical and as much an art as a science.

Notes

1. Kew, National Archives (NA), Harry Ernest Magrath tried at the Old Bailey on 24 May 1880 for rape, CRIM 1/8/2. Recorded as Mcgrath in the Old Bailey Proceedings.
2. For example Louise Jackson's sample of 230 Old Bailey sexual offence cases found that medical witnesses attended every trial; Louise A. Jackson, 'Child Sexual Abuse and the Law: London 1870–1914', unpublished doctoral thesis, Roehampton Institute (1997), p. 166.
3. For some of the many examples of works based on Old Bailey records see Joel Peter Eigen, *Unconscious Crime: Mental Absence and Criminal Responsibility in*

Victorian London (Baltimore; London: Johns Hopkins University Press, 2003); Thomas Rogers Forbes, *Surgeons at the Bailey: English Forensic Medicine to 1878* (New Haven: Yale University Press, 1985); Stephen Landsman, 'One Hundred Years of Rectitude: Medical Witnesses at the Old Bailey, 1717–1817', *Law and History Review* 16 (1998), 445–94; John M. Beattie, 'Garrow and the Detectives: Lawyers and Policemen at the Old Bailey in the Late Eighteenth Century', *Crime, Histoire et Sociétés* 11 (2007), 5–23.

4. London, London Metropolitan Archives (LMA), Pre-Trial Statements, Joseph Harvey tried at the Middlesex Sessions on 28 January 1880 for indecent assault, MJ/SP/E/1880/001.
5. London, LMA, Pre-Trial Statements, Harvey, MJ/SP/E/1880/001.
6. The Social Science Association, for example, existed from 1857 to 1886 and had connections with the British Medical Association throughout this period; see Lawrence Goldman, *Science, Reform, and Politics in Victorian Britain: The Social Science Association, 1857–1886* (Cambridge: Cambridge University Press, 2002). On the emergence of a concept of 'normal' as 'typical' see Ian Hacking, *The Taming of Chance* (Cambridge: Cambridge University Press, 1990), p. 160.
7. Ian Hacking, 'Normal People' in *Modes of Thought: Explorations in Culture and Cognition*, ed. Nancy Torrance and David R. Olson (Cambridge: Cambridge University Press, 1996), 59–71, p. 69.
8. Brian Caddy and Peter Cobb, 'Forensic Science' in *Crime Scene to Court: The Essentials of Forensic Science*, ed. Peter White, 2nd edn (Cambridge: The Royal Society of Chemistry, 2004 [1998]), 1–20, pp. 5–6; William J. Tilstone, Kathleen A. Savage and Leigh A. Clark, *Forensic Science: An Encyclopedia of History, Methods, and Techniques* (Santa Barbara, CA: ABC-CLIO, 2006), p. 28; Jo Lovett, Linda Regan and Liz Kelly, *Sexual Assault Referral Centres: Developing Good Practice and Maximising Potentials*, Home Office Research Study 285 (London: HMSO, 2004), p. 5.
9. Francis Ogston, *Lectures on Medical Jurisprudence* (London: J. & A. Churchill, 1878), p. 91.
10. For example in Alfred Swaine Taylor, *Medical Jurisprudence*, 4th edn (London: J. & A. Churchill, 1852 [1844]).
11. Alfred Swaine Taylor, *The Principles and Practice of Medical Jurisprudence*, ed. Thomas Stevenson, 6th edn, vol. 2 (London: J. & A. Churchill, 1910 [1865]).
12. This statement is drawn not from any specific text(s), but is based on a broad survey of late nineteenth- and early twentieth-century physiological literature in which these three issues were repeatedly represented – sometimes implicitly.
13. Lynn Eaton, 'College Looks Back to Discovery of Hormones', *British Medical Journal (BMJ)*, 25 June 2005, 1466, p. 1466.
14. For example T. S. Clouston, 'Puberty and Adolescence Medico-Psychologically Considered', *Edinburgh Medical Journal* 26 (1880), 5–17, p. 6; *Advice to the Married and Those Contemplating Marriage of Both Sexes: A Book for the People of Valuable Information and Advice, to Teach the Married and Marriageable How to Conduct Themselves*, 2nd edn (London: E. Seale, 1911 [1907]), p. 14; Elizabeth Blackwell, *The Moral Education of the Young in Relation to Sex*, 6th edn (London: Hatchards, 1882 [1878]), p. 9.
15. Sara Read, *Menstruation and the Female Body in Early Modern England* (Basingstoke: Palgrave Macmillan, 2013), p. 47.

16. William B. Carpenter, *Principles of Human Physiology: With Their Chief Applications to Psychology, Pathology, Therapeutics, Hygiene & Forensic Medicine*, 5th edn (London: Blanchard & Lea, 1855 [1842]), p. 792; this was also cited in William Acton, *The Functions and Disorders of the Reproductive Organs in Youth, in Adult Age, and in Advanced Life, Considered in Their Physiological, Social and Psychological Relations* (London: John Churchill, 1857), p. 5; Elizabeth Garrett Anderson, 'Sex in Mind and Education: A Reply', *Fortnightly Review* 15 (1874), 582–94, p. 589.
17. Clouston, 'Puberty and Adolescence', p. 12.
18. For an explicit discussion of the 'parallel' development question see C. Seitz, 'Diseases of Puberty' in *The Diseases of Children: A Work for the Practising Physician*, ed. M. Pfaundler and A. Schlossmann, trans. John Howland, vol. 2 (Philadelphia; London: J. B. Lippincott, 1908), 111–30, p. 114.
19. Literature on secondary sexual characteristics remained largely unchanged from the mid-nineteenth into the early-twentieth century. See Carpenter, *Principles of Human Physiology*, 5th edn, p. 792; G. Stanley Hall, *Adolescence: Its Psychology and Its Relations to Physiology, Anthropology, Sociology, Sex, Crime, Religion and Education*, vol. 1 (London; New York: Appleton, 1904), p. 415.
20. As medical statisticians generally used retrospective interviews, microscopic tests were not possible. On the importance of spermatozoa see Jean L'Esperance, 'Doctors and Women in Nineteenth-Century Society: Sexuality and Role' in *Health Care and Popular Medicine in Nineteenth Century England: Essays in the Social History of Medicine*, ed. John Woodward and David Richards (London: Croom Helm, 1977), 105–27, p. 105; William A. Guy and David Ferrier, *Principles of Forensic Medicine*, 5th edn (London: H. Renshaw, 1881 [1844]), p. 66.
21. The age of menarche was relatively easy to pinpoint and was the most best outward indicator of maturity, although medical writers acknowledged that was not an entirely reliable indicator of the capacity to reproduce; see, for example, H. Macnaughton-Jones, 'The Relation of Puberty and the Menopause to Neurasthenia', *The Lancet*, 29 March 1913, 879–81, p. 880; H. Aubrey Husband, *The Student's Handbook of Forensic Medicine and Medical Police* (Edinburgh; London: E. & S. Livingstone; Simpkin, Marshall & Co., 1874), p. 69. On the significance of blood across time and place, see: Thomas Buckley and Alma Gottlieb, 'A Critical Appraisal of Theories of Menstrual Symbolism' in *Blood Magic: The Anthropology of Menstruation*, ed. Thomas Buckley and Alma Gottlieb (Berkeley, CA; London: University of California Press, 1988), 1–54.
22. For example, the sexologist Havelock Ellis was unusual in using interviews to assess when sexual pleasure emerged. This study related to his general interest in sexual 'instincts' rather than the sexual body, but Ellis only interviewed 12 women; Havelock Ellis, *Studies in the Psychology of Sex*, vol. 3 (Philadelphia: F. A. Davis, 1904 [1903]), p. 170.
23. For a summary of the history of medical surveys of menarche in Britain and the range of average ages that they pinpointed in the period under study, which only varied by minute degrees, see J. M. Tanner, *A History of the Study of Human Growth* (Cambridge: Cambridge University Press, 1981), pp. 288–93. On early modern sexual maturity see Hannah Newton, *The Sick Child in Early Modern England, 1580–1720* (Oxford: Oxford University Press),

p. 8; Sarah Toulalan, '"Unripe" Bodies: Children and Sex in Early Modern England' in *Bodies, Sex and Desire from the Renaissance to the Present*, ed. Kate Fisher and Sarah Toulalan (Basingstoke: Palgrave MacMillan, 2011), 131–50, pp. 136–37.
24. John Roberton, 'On the Alleged Influence of Climate on Female Puberty in Greece', *Edinburgh Medical and Surgical Journal* 62 (1844), 1–22, pp. 1–10; 'Bibliographical Notices: John Roberton, *Essays and Notes on the Physiology and Diseases of Women, and on Practical Midwifery* (London: 1851)', *American Journal of the Medical Sciences* 22 (1851), 179–96, pp. 179–80.
25. Read, *Menstruation and the Female Body*, p. 41; Toulalan, '"Unripe" Bodies', p. 136.
26. John Roberton, *Essays and Notes on the Physiology and Diseases of Women* (London: John Churchill, 1851), p. 30. These statistics had previously been published in essay form.
27. Charles Roberts, 'The Physical Maturity of Women', *The Lancet*, 25 July 1885, 149–50, p. 149.
28. Roberts, 'The Physical Maturity of Women', p. 149.
29. For a receptive discussion of Roberton's findings see Charles Bell, 'The Constitution and Diseases of Women, Part II', *Edinburgh Medical and Surgical Journal* 62 (1844), 311–30, p. 324. Roberts was also cited positively in 'The Criminal Law Amendment Bill', *The Lancet*, 8 August 1885, 252, p. 252.
30. George Black, *The Young Wife's Advice Book: A Guide for Mothers on Health and Self-Management*, 6th edn (London: Ward, Lock & Co., 1888), p. 4.
31. John Roberton's 1851 study, for example, creates a bell curve when converted from table to graph. The Gaussian (Normal) Distribution or bell curve emerged to depict the distribution of heads in coin tossing. It therefore was a purely mathematical concept until Quetelet took the 'average man' as the object of such a statistical study. For a history of the bell curve see Lynn Fendler and Irfan Muzaffar, 'The History of the Bell Curve: Sorting and the Idea of Normal', *Educational Theory* 58 (2008), 70–82. On Quetelet see also: Anna G. Creadick, *Perfectly Average: The Pursuit of Normality in Postwar America* (Amherst; Boston: University of Massachusetts Press, 2010), p. 144; and Lars Grue and Arvid Heiberg, 'Notes on the History of Normality – Reflections on the Work of Quetelet and Galton', *Scandinavian Journal of Disability Research* 8 (2006), 232–46, pp. 234–35.
32. Georges Canguilhem, *The Normal and the Pathological*, trans. Carolyn R. Fawcett (New York: Zone Books, 1989), pp. 154–55.
33. Hacking, 'Normal People', p. 71.
34. 'Precocity, n.', OED Online (Oxford: Oxford University Press, 2015). <http://www.oed.com/view/Entry/149690> (accessed June 22, 2015).
35. Claudia Nelson, *Precocious Children and Childish Adults: Age Inversion in Victorian Literature* (Baltimore: JHU Press, 2012), p. 104.
36. George R. Murray, 'Address in Medicine on some Aspects of Internal Secretion in Disease', *The Lancet*, 26 July 1913, 199–204, p. 204.
37. R. L. Langdon-Down, 'Precocious Development', *BMJ*, 21 September 1907, 743–47, p. 744.
38. Hacking, 'Normal People', p. 71.
39. Patrick Geddes and J. Arthur Thomson, *Sex* (London: Williams and Norgate, 1914), p. 117.

40. Most practitioners were registered after it became a professional requirement in 1858, a process that often included recording their educational and professional backgrounds. The *Medical Directory* shows that Middlesex, Gloucestershire, Devon and Somerset practitioners often had similar medical training in London or Edinburgh and previous work experience in a range of locations. On education see Jennifer Ward, 'Origins and Development of Forensic Medicine and Forensic Science in England 1823–1946', unpublished doctoral thesis, Open University (1993), p. 22.
41. Anne Digby, *Making a Medical Living: Doctors and Patients in the English Market for Medicine, 1720–1911*, paperback edition (Cambridge: Cambridge University Press, 2002 [1994]), p. 12.
42. London, LMA, Pre-Trial Statements, Alfred Boydry tried at the Middlesex Sessions on 20 August 1877 for indecent assault, MJ/SP/E/1877/019.
43. Peter W. J. Bartrip, *Mirror of Medicine: A History of the British Medical Journal* (Oxford: British Medical Journal; Clarendon Press, 1990), p. 9.
44. Bartrip, *Mirror of Medicine*, pp. 11, 185.
45. 'Age of Puberty in England: Mr. Whitehead and Mr. Roberton', *BMJ*, 9 February 1848, 83, p. 83.
46. Roberts, 'The Physical Maturity of Women', pp. 149–50.
47. See, for example, James Milner, 'Note on a Case of Precocious Conception with Subsequent Delivery at Full Term', *The Lancet*, 7 June 1902, 1601–02, p. 1601.
48. Taylor, *Medical Jurisprudence*, 4th edn, pp. 564–65; Taylor, *Principles and Practice*, ed. Thomas Stevenson, 4th edn, vol. 2 (London: J. & A. Churchill, 1894 [1865]), p. 304; Alfred Swaine Taylor, *A Manual of Medical Jurisprudence*, 10th edn (London: J. & A. Churchill, 1879 [1844]), p. 582.
49. This comment was in *Principles and Practice* from 1873 and first appeared in the *(Manual of) Medical Jurisprudence* series in Taylor, *Manual*, 10th edn, p. 649.
50. This comment was in the first edition of Taylor, *Principles and Practice*, vol. 1 (London: J. & A. Churchill, 1865), p. 18, and was repeated in subsequent editions.
51. H. A. Lediard, 'Our Criminal Procedure: Scientific Imperfections and Remedy', *The Lancet*, 18 October 1890, 812–14, p. 813.
52. '*Medical Jurisprudence* by Alfred Swaine Taylor, M.D., F.R.S., F.R.C.P. Seventh Edition', *BMJ*, 9 November 1861, 505, p. 505.
53. *Old Bailey Proceedings Online* (*OBPO*) (www.oldbaileyonline.org, version 7.0), trial of William Slater William Vivian in September 1860, t18600917-769; *OBPO*, trial of Hannah Maria Pipkins in January 1855, t18550129-261.
54. London, LMA, Pre-Trial Statements, James Westhall tried at the Middlesex Sessions on 29 September 1868 for carnal knowledge, MJ/SP/E/1868/017; Gloucester, Gloucestershire Archives (GA), Pre-Trial Statements, Thomas Cobb tried at the Gloucestershire Quarter Sessions on 19 October 1898 for assault with intent, Q/SD/2/1898.
55. '*Principles of Forensic Medicine* by W. A. Guy, M.B. Cantab., Professor of Forensic Medicine, King's College. 3rd edition, enlarged. London: 1868', *Edinburgh Medical Journal* 14 (1868–69), 157, p. 157; '*Medical Jurisprudence* by Alfred Swaine Taylor', p. 505.

56. Exeter, Devon Record Office (DRO), Thomas Bright tried at the Devon Quarter Sessions on 29 June 1876 for assault with intent, QS/B/1876/Midsummer.
57. Nancy M. Theriot, 'Negotiating Illness: Doctors, Patients and Families in the Nineteenth Century', *Journal of the History of the Behavioral Sciences* 37 (2001), 349–68; Willemijn Ruberg, 'Trauma, Body, and Mind: Forensic Medicine in Nineteenth-Century Dutch Rape Cases', *Journal of the History of Sexuality* 22 (2013), 85–104.
58. Gloucester, GA, Pre-Trial Statements, Cobb, Q/SD/2/1898; London, LMA, Pre-Trial Statements, William Roon tried at the Middlesex Sessions on 2 December 1873 for attempted carnal knowledge, MJ/SP/E/1873/027.
59. John Byers, 'Quackery – with Special Reference to Female Complaints', *BMJ*, 27 May 1911, 1239–42, p. 1240.
60. Byers, 'Quackery', pp. 1240–41.
61. Shani D'Cruze and Louise A. Jackson, *Women, Crime and Justice in England since 1660* (Basingstoke; New York: Palgrave Macmillan, 2009), p. 12; Louise A. Jackson, 'Family, Community and the Regulation of Child Sexual Abuse: London, 1870–1914' in *Childhood in Question: Children, Parents and the State*, ed. A. Fletcher and S. Hussey (Manchester: Manchester University Press, 1999), 133–51, p. 134.
62. This quote is taken from one of the very few pre-trial statements with questions recorded; Gloucester, GA, Pre-Trial Statements, Peter Learly tried at the Gloucestershire Quarter Sessions on 27 June 1871 for indecent assault, Q/SD/2/1871. The nature of this pre-trial statement indicates that others re-framed 'question and answer' formats into full sentences, although witnesses' answers were often more extensive than just 'yes' or 'no'.
63. For example *Old Bailey Proceedings Online* (*OBPO*) (www.oldbaileyonline.org, version 7.0), trial of Thomas Mercer in August 1694, t16940830-9; *OBPO*, trial of Edward Hatfield in October 1777, t17771015-10.
64. London, LMA, Pre-Trial Statements, Dorothy Lund and Albert Warcup tried at the Middlesex Sessions on 22 June 1886 for indecent assault, MJ/SPE/1886/031; London, LMA, Pre-Trial Statements, William Moore tried at the Middlesex Sessions on 15 September 1886 for indecent assault, MJ/SP/E/1886/044; *OBPO*, trial of Benjamin Hullock in August 1723, t17230828-64.

2
Injury: Signs and the Sexual Body

In his 1878 *Lectures on Medical Jurisprudence*, prominent medico-legal author Francis Ogston referred to 'Physical Proofs' as the only kind of evidence that came 'properly ... within the province of the medical jurist'.[1] Although medical witnesses spoke about issues related to behaviour and character implicitly, and occasionally explicitly, 'physical proofs' underpinned most medical expertise in trials for sexual offences. In Middlesex, Gloucestershire, Somerset and Devon, medical witnesses generally focused on bodily signs rather than symptoms such as pain.[2] This focus was in part due to the limited ability of young complainants to describe symptoms and the consequence of a general turn towards treating the patient's body as 'object' rather than the patient as 'subject'.[3] This is not to say that the patient's voice was completely removed from all clinical encounters. However, symptoms were generally superfluous to the medico-legal 'script', which followed particular lines of enquiry deemed legally relevant.

Medical witnesses in court interpreted each bodily sign in the light of a developmental framework that drew upon wider physiological theory of the Victorian and Edwardian periods. Drawing upon research that showed the lengthy and variable nature of puberty, this somatic (bodily) developmental paradigm conceptualised the young body as inherently pure, but as gradually losing this purity with age. The approach was a product of its time, combining a new medical interest in developmental studies with Romantic notions of childhood innocence and the legacy of an Evangelical revival concerned about the corruption of youth. Medical focus on the body was only possible because of medical adherence to traditional diagnostic methods. Medical witnesses interpreted signs such as genital bleeding or vaginal discharge in relation to puberty rather than, as would become the case later in the twentieth century, using microscopic analysis to identify menstrual blood or bacteriological methods to identify disease.[4] This knowledge framework relied upon knowledge of physical maturity and implicitly allowed for value judgements if complainants did not fit normative standards of childhood, femininity and masculinity, or 'victimhood'.

Much medical testimony about the abused body focused on the subject of genital violence, in line with the general medico-legal focus on 'physical proofs'. Although in theory a straightforward form of medical evidence, as it was based on describing injury rather than a complainant's character or behaviour, this testimony was also imbued with social concerns. As sociologist Nick Lee notes, 'unmediated' expert evidence has always been a carefully constructed concept; successful medical witnesses ostensibly allowed 'the physical trace to "speak for itself" ', but in practice their testimony was always framed by social, legal and individual concerns.[5] Even the interpretation of 'physical proofs' was not a straightforward scientific process disconnected from wider cultural concerns about age, gender, class and race. This chapter demonstrates the moral and age-based dimensions of 'physical proofs', by considering how ideas about the injuries caused by sex changed with the age of a complainant, then turns to medical evidence about a number of specific injuries: inflammation; bloodstains; pregnancy; and disease.

Most medical testimony about genital violence focused on the female body. Injuries from penetrative sex were expected to include, to quote the London physician Alfred Lee's 1882 pre-trial statement, 'abrasions ... laceration ... dilation of the vagina ... rupture of the labia ... blood stains ... inflammatory symptoms'.[6] The interpretation of such injuries varied according to the age of the girl, but the same signs were cited regularly in relation to females of all ages. The limited discussion of male genital injury related not only to the overall rarity of cases involving males, but also to the expectation that the male body would display fewer signs of a sexual crime than the female body. The indecent touching of a man or boy apparently would not leave any lasting mark unless a venereal disease was transmitted, whereas the same offence committed upon a girl could cause some genital damage if a manual assault was also penetrative.

When males were complainants, only cases involving anal penetration of pre-pubescent boys would involve notable genital injury. According to Alfred Swaine Taylor's influential *Manual of Medical Jurisprudence*, claims made by males were 'commonly sufficiently proved without medical evidence, except in the cases of young persons, when marks of physical violence will in general be sufficiently apparent'.[7] These 'marks of physical violence' on the genital area of young males were similar to those cited for females because, as Louise Jackson has observed regarding the girl's sexual organs, a male child's anus was expected to be naturally 'small and enclosed'.[8] It would therefore be examined in the same way as a young girl's vagina, for signs of recent violence such as laceration (tearing), bruising and blood. From 1874 onwards Taylor also noted the signs of 'one long habituated to ... unnatural practices' in males, in order to distinguish them from the signs of violence, as 'a funnel-shaped state of the parts ... with the appearance of dilation ... and a destruction of the folded' or puckered state of skin in this part'.[9] In other words, the anus of a boy who had regular experience of same-sex intercourse would be shaped more to readily admit penetration. It would be more open, and the skin more smooth, than that of a boy who had not previously experienced such an act. Genital injuries were evidently relevant to trials involving male complainants. However, such evidence was limited to cases of alleged penetrative assaults by adult men on young boys, which were very rare in the Middlesex Sessions and south-west Quarter Sessions. 87 per cent of pre-trial statements from Middlesex and 97 per cent from Gloucestershire, Somerset and Devon involved female complainants.

In relation to female complainants, the nature of medical testimony on genital signs shifted with the age of an alleged victim. Medical witnesses demonstrated a belief that an female infant's body was so small and enclosed as to be asexual. The notion of a child's body as 'unripe' and unready for sex had a long history, reinforced by peculiarly Victorian and Edwardian middle-class concerns about precocity and childhood innocence. The asexuality of youth was embodied by a general smallness of the vaginal entrance and vagina that prevented complete penetration. In infancy, the vagina itself was so small that it was thought by some even to prevent rape. The case of *Rex v. Russen* in 1777 had established that the smallest degree of penetration constituted rape, including those cases in which an infant or child's vagina was too narrow to allow full penetration; this provision was established in statute in 1828, a law that also removed the need for 'emission of seed' to prove the offence.[10] However, a controversy around the question of whether infants' bodies could be raped – seemingly grounded in older definitions of rape that required full penetration – lingered into the early-twentieth century. Throughout all editions of his medical jurisprudence textbooks, Taylor put forward a variation of the following point (this quote taken from his 1910 work): 'It has been questioned whether a rape can be perpetrated on children of tender age by an adult man; and medical witnesses at trials have given conflicting opinions'.[11] Evidence from the courts supports his claims. Of the many cases of sexual crime tried at the Middlesex Sessions in 1869, a case from July is particularly noteworthy for the youth of the alleged victim – a female infant aged just one year and eight months. A charge of indecent assault rather than rape was taken for the trial, in part because a medical witness emphasised that infants could not be raped. In this case, police surgeon George Bagster Phillips declared that 'it would be impossible for the male organ of a man to pass up the child's person'.[12]

Controversies about whether rape was possible ebbed for cases involving older children, but did not disappear. Some medical witnesses emphasised that all young girls were impossible to rape, even after infancy. Middlesex police surgeon Thomas Jackman argued in 1881 that '[i]t would be impossible for a man to have penetrated into the vagina' of a five-year-old girl.[13] This comment echoed some contemporary thought, such as that of French medical jurist Léon Henri Thoinot who claimed as late as 1911 – perhaps drawing upon traditional models of the life cycle constructed in seven-year stages – that 'under six years of age ... a child cannot be raped: the penis cannot enter the internal genital organs'.[14] However, English medical literature more commonly noted that childhood was not a static life stage; in line with developmental studies it was thought that girls' bodies, including the sexual body, changed even in the first few years of life.

Most medical texts and witnesses agreed with Taylor that even full penetration became broadly possible after infancy, but that a rape would cause significant marks of violence due to the relatively – although no longer entirely – small and enclosed nature of such girls' sexual bodies. Francis Ogston, whose book was another 'standard authority' on forensic medicine, noted that genital injury was expected to be most clear in the youngest complainants because of the 'disproportion between the adult penis and impuberant female genitals'.[15] With the exception of Jackman, medical witnesses also took this line in court. In 1874, for example, a Gloucestershire medical witness testified that 'without a dreadful destruction of the parts it would be impossible for a full grown man to have connexion with a girl of five years old'.[16] Medical witnesses in both locations made such claims, about the impossibility of a penetrative sexual assault on young girls without significant physical harm, in relation to girls aged from five to ten years old. Physical harm was thought to be especially likely if – as was the case in Cheltenham in 1898 – a medical witness testified that a young complainant was 'not so fully developed as girls of that age generally are'; such girls, who were more childlike than the norm, embodied the characteristics of victimhood and stood in opposition to the precocious child.[17]

There was no clear or consistent turning point at which the enclosed and pure body changed to a more open and naturally sexual body, but puberty was believed to bring with it significant physical change. Only at puberty was it thought that the vaginal entrance and vagina would open up sufficiently to allow for penetration without great physical harm, although signs of injury were still expected if a girl was a virgin. In the words of the *British Medical Journal* of 1903, 'the vagina was wide ... in a woman in the ages of sexual maturity'.[18] Helen King's work on ancient Greece shows that ideas about the gradual opening up of female bodies at puberty had a long history, with early medical ideas about puberty focusing on the 'widening of ... channels so that [humoral] fluids can move around more easily'.[19] The physiological explanations had changed by the nineteenth century, for example with a focus on changes to the vagina's bacterial and cell composition rather than humoral fluids, but the concept of the body 'opening up' at puberty was not a new one. In consequence of the gradual widening of the vagina it was thought that, in Ogston's words, 'where both parties are above puberty, it is obvious that the same disproportion [of genital parts] may still exist, though in a less degree, and produce effects in the female proportionate in degree, though the same in kind'.[20]

The assumption that a lesser degree of harm would occur to a girl nearing or at the age of puberty, when her body was no longer inherently pure and enclosed, influenced medical witness testimony in court. As puberty apparently could occur in healthy girls at a wide age range, medical witnesses became increasingly reluctant to make any assumptions about sexual maturity for girls over the age of ten. In 1860 a medical witness in Somerset examined a girl aged 11 in order to evaluate her level of development, rather than assuming that she was (not) sexually mature, and testified that '[t]here are marks of violence about the [complainant's] private parts and I have no doubt some person has had very recent connexion with her ... Her parts are not fully developed'.[21] As she was not 'fully developed', marks of violence were necessary to prove an offence. Similarly, a Devon medical witness in 1885 treated the 'impossibility' of sexual intercourse with an 11-year-old girl as a question to be evaluated rather than assumed; he only declared that 'penetration was impossible without laceration' after examining the 'size of the parts'.[22] The prisoner admitted his crime, but on the basis of this evidence was convicted on the lesser charge of indecent assault. Genital injury could indicate violence at any age, but its absence in the young or undeveloped body was particularly noteworthy as a sign because it was thought to be a certain indicator that penetration had not occurred.

These medical witnesses gave meaning to the absence of injury by highlighting the girls' small and enclosed – thus asexual – bodies, but only after examinations of the complainants. Aged 11, these girls were at the cusp of the 'normal' age range for puberty; if found to have reached puberty, with its associated opening up of the body and natural genital dilation, the absence of signs of injury would have been less significant as evidence. The asexuality of girls at or close to the age of puberty was determined on a case-by-case basis. Rather than assuming that a girl was not sexually mature on the basis of her age alone, as in sexual consent law, medical witnesses approached bodily norms as flexible. They evaluated sexual capacity as girls neared the age of possible puberty, but deemed no such examination necessary for infants or married women whose respective asexuality and sexual maturity could be assumed. Overall, medical men emphasised that, due to changes in the female body: rape of infants was unusual or even impossible; a penetrative assault on young girls would result in extensive and obvious injury such as laceration; and older girls could be assaulted with little or even no injury, unless a particularly violent assault. In theory, this framework of evidence meant that pre-pubescent female bodies were the easiest to interpret. The younger the body, the more pure it was thought to be, and the clearer the meaning of the presence or absence of injury.

These claims only applied to cases of penetrative sexual crime that were actually relatively rare in mid-level courts. More often, medical witnesses needed to interpret signs such as inflammation and redness, rather than bruising and laceration, that were ambiguous for girls of any age. For young girls, a range of natural causes could explain inflammation. In a case of alleged assault on a girl aged nine from 1857, for example, a Middlesex medical witness noted under cross-examination that 'heat of stomach or even walking would produce inflammation in so young a child'.[23] In such cases bodily heat provided a possible 'natural' but pathological explanation for potential indicators of genital harm, such as redness and inflammation. In another case from Middlesex, tried in 1883, a complainant's grandmother noted that she had observed that the eight-year-old girl was 'raw' but that she made no criminal complaint because 'I thought the rawness was from the heat of her body'.[24] In this case the girl's neighbour made a similar statement, stating that she 'found the parts very sore ... I thought it was from the over heating of her blood'.[25] These claims were similar but not identical.

'Heat of stomach' was a medical diagnosis prevalent in the literature of the early-nineteenth century, largely in relation to digestive disorders, while ideas about heat of blood related more to concerns about fever. Despite these differences, overlaps between the two testimonies shows interactions between lay and medical knowledge in relation to traditional models of bodily 'heat' as pathological. 'Heat' was just one, of many, possible explanations for genital inflammation in the young.

Not only were medical witnesses often unsure about the meaning of signs such as inflammation, but the court system also encouraged ambiguity. Medical witnesses, more than any other kind, were pressed under cross-examination on whether they would be prepared to 'swear' to their testimony. This defence strategy relied on the fact that medical witnesses were generally reluctant to do so, both for professional reasons and because of the limitations of diagnostic methods. In a trial from Devon in 1876 for an alleged offence on a 'little girl', for example, the medical witness was pressed once by the prosecution and twice by the defence to 'swear' to his opinion. He testified that:

> I am a Surgeon residing at Ashburton. On Thursday the 3rd August instant I saw Alice Bradford. I examined her on that day. She was then on the bed in her mother's house. I saw that her external genitals were red and a little raw looking from some cause or other. I cannot swear from what cause the redness arose. The little girl told me from what it arose. I can account for the redness from what the girl told me. (cross-examined) I did not see it until 4 days afterwards ... I cannot swear that the redness was caused by unnatural means. I cannot swear from what it arose.[26]

The vague nature of this testimony made it extremely useful for the prisoner's defence. The surgeon refused three times to 'swear' to the cause of the girl's redness and noted in the broadest terms that it was from 'some cause or other'. This evidence acknowledged the ambiguity of signs on a child's body, with a sign such as inflammation mapping onto meanings as varied as sexual assault and common disease. The ambiguous nature of these signs must have contributed to the trial outcome, where it was ruled that there was no evidence to go to the jury. The prisoner was simply sent away with a warning 'to be cautious not to interfere with and insult young girls in future'.[27]

The problems surrounding bodily norms fuelled such ambiguity and opened up space for moral questions in the courtroom, particularly those around precocity and age-appropriate behaviour or appearance.

When Elizabeth Thorne, aged 11, complained of an assault in 1869, her grandmother took her to a Middlesex police station where a medical practitioner made an examination. The girl testified that 'What he pulled out of his trousers he called a "Cock". He put "it" to my "Diddle" – between my legs, it went inside – and remained in five or ten minutes he hurt me very much'.[28] The first medical witness, a surgeon, examined Thorne for signs of a penetrative sexual assault and found inflammation and bloodstains. During initial questioning in the magistrates' court, he stated a strong belief that such marks were signs of partial penetration. However, under cross-examination he admitted that '[i]t is not common, but it does happen that children's parts are inflamed often without any external violence. The girl's parts were not very small for her age'.[29] The medical practitioner wavered on his earlier conviction under questioning, acknowledging that a young girl's body might differ from the norm 'for her age' without being the result of an assault. The prisoner's defence elicited this medical testimony because it raised questions about her status as victim. This girl was on the brink of possible puberty and was, according to the medical witness, more sexually developed than her peers. Signs that would have been certain indicators of assault in a 'normal', small and enclosed child's body lost some significance with such testimony. A different medical witness in the same trial, called by the defence, declared that 'drops of blood on the child's linen might be from natural causes. Some children menstruate before they are 12 years old'.[30] This witness drew his knowledge of sexual maturity not only from medical literature but also, he claimed, from 'immense experience in Women's Cases'.[31] Although the surgeon gave clear opinions within his testimony, being 'prepared to swear' to his opinion that no penetration had occurred, he also acknowledged the ambiguity of menarche.[32] The prisoner's defence used the variable nature of sexual maturity to place doubt on the link between bloodstains and violence, with success: the prisoner was acquitted.

This case raises the issue of bloodstains as a sign, which carried great significance because of the symbolic importance of menstruation as a marker of maturity. As in the Thorne case, medical witnesses often gave meaning to genital bleeding by considering typical ages of menarche. Bleeding and bloodstains as signs were particularly closely interwoven with medical knowledge about ages and stages of sexual development, as they became potentially 'natural' and 'normal' at puberty. Victorian forensic medicine textbooks proposed some techniques to differentiate menstrual from non-menstrual blood microscopically but, due to a lack of resources and limited hands-on training in these new methods, most general practitioners used naked-eye diagnosis into the early-twentieth century.[33] Without microscopic methods, medical witnesses needed to draw upon more ambiguous knowledge about 'normal' ages of puberty in order to interpret bodily signs. In 1903 a Somerset general practitioner testified under cross-examination that stains on the drawers of an eight-year-old girl 'might have resulted from natural causes but it is extremely unlikely. I have never seen a girl of that age suffering from her courses'.[34] This testimony showed that bodily norms were flexible, but were not infinitely so; medical witnesses trod a fine line between emphasising the ambiguities and flexible nature of sexual maturity, and noting that certain bodily signs were unlikely at a given age. The medical witnesses in this case did not dismiss menstruation at the age of eight as entirely impossible, but indicated that it would not be the norm. His testimony created space for doubt around the meaning of bloodstains, encouraged by defence questioning, but did not present early menarche as a likely explanation for bloodstains found on linen.

If a girl was found to have reached menarche or gained reproductive capacities prematurely, or precociously, her status as victim often came into question. Concerns about precocious puberty were implicitly widespread in the courts, but were articulated most explicitly in relation to concerns about social class and the unusual but important question of ethnicity. Although ethnic diversity was limited in the Middlesex Sessions and Quarter Sessions, the legal implication of links between race and precocity was evident in other contexts. Concerns about the future of the British Empire and the growing eugenics movement fuelled anxieties about race-based precocity in the early-twentieth century, both within and beyond the medical process. In a case from the Court of Criminal Appeal from 1913, for example, the 12-year-old female complainant of an unspecified 'different race' was deemed to be sexually precocious. The appeal judge stated that:

> The appellant was convicted under s. 4 of the Criminal Law Amendment Act, 1885. All cases of that kind are very grave, but it must be remembered that the people concerned in this case are of a race which develop at an earlier age than English people ... Both the man and the girl are of a different race to the people judging them and applying the law of England to them. Trying not to be rendered indignant by the fact that the girl was rendered pregnant, which was not the offence, we feel that the justice of the case would be met by a sentence of five years' penal servitude. Sentence reduced.[35]

This judge thus set a lower penalty for carnal knowledge of girls of a different 'race', on the basis of their apparently early sexual maturity.

Although the girl's pregnancy was considered to be a 'grave' outcome of the crime, which had led the original judge to be 'rendered indignant', the fact that she was capable of being impregnated also worked as evidence against her by proving her physical precocity. This complainant was, however, only precocious by Western European standards and apparently fitted the norms of her racial background.

Ideas about race-based precocity overlapped with an equally widespread belief in links between climate and precocity. Females from hotter climates or working-class 'factory girls' who worked in hot environments were expected to develop earlier and to be more sexually active.[36] In 1898, for example, *The Lancet* reported that '"[t]he Englishman and the Scandinavian," says Ferrero, "are sexually less precocious and cooler-blooded than the Frenchman or the Spaniard"'.[37] Although medical statistician John Roberton showed that climate was not an influential factor on age of menarche, heat and sexuality remained inextricably woven in medical and lay thought.[38] These medical theories on heat presented in court were closer to religious and popular thought than to cutting-edge science. They had echoes of older Galenic theories about the link between bodily warmth and 'ripening' in which puberty was precipitated by a rise in bodily heat, which particularly disrupted the female body as it was typically deemed to be colder than that of males.[39] Bodily heat, linked to menarche, was apparently typical for girls at 'normal' or precocious puberty and was interwoven with value judgements about sexual behaviour. Links between physical and behavioural precocity explain general medical concerns about precocious menarche and uncontrolled sexuality. Physician George R. Drysdale even compared this sexual heat to the 'period of heat in female animals' in his bestselling work *The Elements of Social Science*.[40] Bodily 'heat' had been a physical pathology for the very young, but was potentially a moral pathology for precocious and pubescent girls.

New medical research on sexual maturity, which had shown that girls as young as 12 could fall in the 'normal' age range for menarche, seemingly did not mitigate moral concerns about the sexual precocity of girls at these ages. Links between physical sexual maturity and sexual behaviour meant that medical witnesses and writers continued to vocalise concerns about girls who reached puberty earlier than average. Only once a complainant reached the age of 16 was there a widespread expectation that they would have reached menarche and that bleeding might be 'natural'. Although the disrupting influence of 'heat' of body on sexual behaviour was still a concern in such cases, the 'normal' and 'natural' timing of such development meant that menarche apparently aligned more closely with mental capacities for modesty and self-control.

The tone of sexual forensics shifted from menarche being possible from around 11–14 to implicitly likely for this older age group, as medical witnesses made statements such as '[t]he blood might have been caused by a finger being inserted, or by her courses' and '[t]he red stain on her drawers may have been from menstrual discharge'.[41] However, the relationship between age and menarche was never straightforward, not least because medical writers and witnesses alike emphasised that puberty was a variable process.

Other signs of harm, such as venereal disease, were theoretically less age-dependent than bleeding, inflammation and other genital injury. Although the communication of disease could occur at any age, it remained important for medical practitioners to consider the 'natural' and 'normal' genital discharges of different life stages in order to differentiate them from signs of disease. Differing from other bodily markers already discussed, medical witnesses also expressed uncertainty about the very nature of genital discharges as signs. Despite scientific developments in the microscopic and chemical interpretation of disease during the late-Victorian and Edwardian periods, medical witnesses were genuinely uncertain about the value of new bacteriological diagnostic methods and about how to distinguish venereal from non-venereal diseases. Medical witnesses often continued to use traditional diagnostic methods, which were highly interpretive in nature and drew upon ideas about the diseases associated with particular life stages. Physiology, which emphasised the variability of those life stages, fuelled uncertainty about the diagnosis of venereal diseases while bacteriology did little to resolve it. Medical witnesses also brought middle-class anxieties about class, contamination and the dangers of puberty to bear on the diagnostic process. As these concerns were shared with judges and jurors, uncertainty did not destabilise medical authority as feared. Like the more general ambiguities of sexual maturity and bodily norms, medical uncertainty about diagnosing disease operated to open up space for interpretation and for the moral aspects of medical testimony.

Doctors were alert to the possibility of finding either gonorrhoea or syphilis in suspected criminal cases, but Gram staining (1884) and Wassermann testing (1906) were not widely used. The absence of Gram staining is particularly significant. Gonorrhoea was the most common venereal disease diagnosed in court cases, but Gram staining was not used once in the Middlesex or south-west courts in the 30 years following its discovery.[42] Most medical practitioners relied heavily on naked-eye diagnosis to diagnose venereal disease throughout the period, drawing upon the same methods that they had used before bacteriological testing was possible. One reason for medical witnesses' delayed adoption of Gram staining can be found by examining books of medical jurisprudence, which guided testimony in court and were also slow to respond to diagnostic shifts. Gradual changes between editions of Taylor's work demonstrate the hesitant integration of bacteriological diagnostic techniques into medical jurisprudence textbooks for practitioners. It was not until 1905 that Taylor's *Principles and Practice of Medical Jurisprudence*, which had been updated by Frederick J. Smith, gave detailed instructions for Gram staining and explicitly stated its faith in the process.[43] There was thus a 21-year gap between the invention of Gram staining and its inclusion as a recommended process in the most major forensic textbook of the period.

Although quicker than medical jurisprudence texts in discussing new developments, *The Lancet* and *British Medical Journal* also reported many of their associated problems. Such journals referred to ongoing concerns about the unreliability of the Gram staining process, particularly in terms of the possibility that another non-gonorrhoeal micrococcus could be mistakenly identified as the gonococcus.[44] Medical readers and witnesses would not have been completely ignorant of the development of Gram staining processes, but were reading journals and jurisprudence texts in which their reliability remained unconfirmed. Such anxieties about the unreliability of Gram staining had faded by the turn of the century.[45] However, the court cases from Middlesex and south-west England indicate that this slow growth of support for Gram staining in medical literature was too late to influence general practice before 1914. The continued emphasis on naked-eye techniques fed into and supported the growing age-based uncertainty of medical testimony in court. There were universal practical limitations to naked-eye diagnosis, but it is significant that medical witnesses did not apply a language of uncertainty consistently. For girls aged between four and eight years old, for example, medical witnesses – across all cases – diagnosed around a third of genital discharges as venereal disease. They were unable clearly to diagnose a third of cases and interpreted the final third as a combination of natural causes and violence without disease. For girls over the age of 16, in contrast, not a single medical practitioner diagnosed a genital discharge as a venereal disease; medical witnesses were also unable clearly to interpret the discharge in half of these cases.

Uncertainty about the diagnosis of venereal disease was ever-present but differed according to the age of a complainant. Medical witnesses explained this uncertainty in terms of the 'natural' diseases and discharges that were common at different life stages. They claimed, for example, that diagnostic uncertainties about disease in young children stemmed from the difficulties of distinguishing venereal disease from commonplace childhood diseases such as leucorrhoea ('the whites') and vaginitis. The age-specific nature of such diseases is indicated in medical statements in court such as, 'female children after fever frequently have such a discharge from their genitals' and 'it is common in young children to have leucorrhoea discharge'.[46] Books of medical jurisprudence, such as those by Francis Ogston, also commented that females under the age of puberty were particularly subject to 'diseases of the genitals naturally occurring .... [N]ot unfrequently encountered in the female children of scrofulous parents, especially amongst the poor and uncleanly'.[47] His work draws attention to the class-based nature of these concerns, which was also evident in medical testimony that attributed discharge in young girls to causes such as 'worms' and 'not being well nourished'.[48]

A general belief in the loss of inherent bodily innocence at puberty also influenced medical testimony about vaginal discharges. In cases involving older complainants, medical uncertainty was grounded in the belief that healthy vaginal secretions became commonplace after sexual maturity. Although medical witnesses emphasised that naturally-occurring disease was common in young girls of the working classes, this physical state was atypical, unhealthy and undesirable. In contrast, as girls grew older the lack of purity of the sexual body was no longer assumed. In one court case involving a 17-year-old complainant the medical witness referred to the visual difference between 'seminal discharge' and 'the ordinary discharge I should expect to find' in a girl of that age.[49] Medical diagnosis shifted gradually in tone with the age of a complainant because puberty was a lengthy and variable process, therefore there was no single age at which vaginal discharges shifted from pathological to potentially healthy and 'ordinary'.

Medical theories about diseases associated with particular life stages had some overlaps with general knowledge about so-called 'female complaints'. Quack literature of the mid-nineteenth century likewise emphasised that '[e]very age has its peculiar affections', although unlike mainstream medical texts often continued explicitly to explain these physical symptoms, discharges and diseases in humoral terms.[50] Medical practitioners also drew upon their professional 'experience in cases of gonorrhoea', in the words of one medical witness from Middlesex, and discussed the naked-eye interpretation of genital discharges with mothers of young children in the clinic.[51] Mothers often brought cases to medical attention when their children had a discharge that differed from the expected diseases of childhood.[52] There was a shared pool of medical and lay knowledge about the signs of healthy and unhealthy (or 'normal' and 'abnormal') children's bodies.

Even if medical witnesses were able to diagnose a venereal disease, on the basis of their knowledge and experience, they then faced the question of whether it was transmitted by a sexual offence. Medical witnesses only identified venereal disease in cases involving girls under the age of sexual consent, in which there was no legal possibility that the venereal disease had been caught from consensual sexual activity. Their testimony was therefore usually corroborative of a criminal charge, as doctors generally agreed with an 1888 *British Medical Journal* article that 'if the [girl's] discharge is gonorrhoea, there would be a strong presumption that it was contracted in the usual way'.[53] Medical jurisprudence texts noted that venereal disease could be transmitted from sponges or water closets.[54] However, only three medical practitioners from all regions in the period 1850–1914 mentioned the possibility of transmission by non-sexual contact in court, none of whom ever denied the possibility that the venereal disease corroborated the charge. At Hammersmith Police Court in 1868, Joseph Smith declared that 'it is stated in Books of Medical Jurisprudence that gonorrhoea might be communicated at the closet. I consider this may be the case' but also commented that 'it might be communicated by contact'.[55] Despite occasionally acknowledging the possibility of alternative modes of transmissions, medical witnesses were generally prepared to interpret a venereal disease as corroborative evidence for alleged sexual offences against children.

This general medical acceptance of the possibility and prevalence of criminal sexual offences against children was part of wider ideas about childhood and child protection. By 1873, many women's rights campaigners and medical practitioners had apparently 'buried the hatchet' about the controversial Contagious Diseases Acts in order to promote the raising of the age of sexual consent.[56] This unusual alignment was part of a widespread emphasis on the true extent of sexual offences against children, as particularly propagated by W. T. Stead's revelations about the 'white slave trade' in young girls in the *Pall Mall Gazette*.[57] Medical practitioners' apparent preparedness to diagnose venereal disease in younger children and to attribute them to violence must be understood within this specific social context, particularly as medical testimony would have been received by a public jury who were likely to have been acquainted with media coverage of such cases.

The English courts provide many points of comparison with Gayle Davis's work on early twentieth-century Scottish asylums, in which she has found that physicians' use of Wassermann testing for syphilis was 'mediated by a range of institutional, professional and social factors as well as the scientific, technical and instrumental features of the procedure'.[58] However, the English context differed from Scotland in other respects. Roger Davidson has demonstrated that a degree of 'denial' existed amongst Scottish medical practitioners in the early-twentieth century, in terms of some reluctance to diagnose venereal disease in children and to link them to sexual contact.[59] Conversely in the English legal system, albeit in a slightly earlier time period, medical practitioners were actually *more* likely to diagnose venereal disease in cases involving girls under the age of 16 and to attribute such diseases to sexual crime. Although English and Scottish doctors apparently used comparable discursive frameworks of 'disease, dirt and pollution' in cases involving the youngest complainants, English doctors never used such concepts to deny that sexual crimes were committed against children.[60] This absence of 'denial' could be linked to the specifically English concerns of the late-nineteenth century, in which the printed media and early feminists emphasised the reality of a 'white slave trade' and the widespread incidence of sexual crimes against children.

In theory, due to the rise of bacteriological diagnostic methods, the diagnosis of venereal disease was a much more clinical and scientific subject than other medical testimony in court. However, it is clear that testimony on this form of genital sign was not a straightforward process. Medical witnesses drew upon interactions with laypeople and wider professional and physiological concerns in order to interpret genital discharges in girls. There were universal difficulties in using naked-eye diagnosis to interpret vaginal secretions, but witnesses articulated greater uncertainty about the interpretation of such signs in older girls. Possible explanations for this trend include physiological theories about the ambiguous and variable nature of sexual maturity, and wider social contexts that emphasised the prevalence of crimes against children. Medical witnesses retained a notion that a healthy child would be physically pure while articulating anxieties about the contaminated lower classes. They also understood and interpreted the sexual body through a somatic developmental framework, in which bodily change was a gradual process, rather than interpreting bodily signs in line with the two-tier legislative system.

Although genital injuries took a wide range of forms, with a complainant's age having slightly different implications for each, in general medical witnesses emphasised that the possible meanings of such signs multiplied as girls neared the age of puberty. Such testimony aligned more with the physiological emphasis on ambiguity than with the law on sexual crime, but this is not to say that sexual forensics had no impact in the courts. Conversely, judges and juries were seemingly receptive to the ambiguities of medical testimony about pubescent complainants, which created valuable space for interpretation. Although the legislative system adhered to relatively rigid ages of sexual consent, the courts often aligned implicitly with medical thought on the subject of genital injury.

It is difficult to know the degree to which medical evidence on genital injury informed magistrates' decisions to dismiss cases or try them summarily, due to the limited survival – in manuscript or printed form – of records for cases that did not reach trial. There is, however, more extensive evidence relating to the role of magistrates in (re)shaping the charge under which a prisoner was tried: magistrates in Middlesex and the South West seemingly downgraded some charges from felony to misdemeanour before passing them forward for trial. The charge was connected directly to the subject of genital violence, as penetration of the vulva was necessary to prove a charge of carnal knowledge or rape in English law. However, the influence of medical testimony on charges was not straightforward. Misdemeanour cases in which medical witnesses made statements such as that 'I believe there had been penetration of the vulva ... but not far enough to rupture the hymen' or 'I found that there had been a certain amount of penetration' should technically have been tried as the felonies of rape or carnal knowledge of girls in the felony age category.[61] Instead, they were often committed to the Middlesex Sessions or county Quarter Sessions as misdemeanours such as indecent assault or attempted carnal knowledge or rape. The reasons for such discrepancies between medical testimony and criminal charges are not entirely clear. Although Louise Jackson claims that surgeons 'tended to argue for lesser charges of indecent or common assault', there is no definitive evidence that any medical witnesses explicitly advocated any particular charge.[62] Medical practitioners were not always even aware of the legal repercussions of their testimony, as indicated by a letter to *The Lancet* editor in 1843 that asked about the degree of penetration required to constitute a rape in law.[63] The crucial question is arguably not whether medical practitioners 'argued for' lesser charges, but why medical depositions about partial penetration were disregarded despite being elicited in court.

It is impossible fully to reconstruct the motives of magistrates, but there are some possible explanations for their decisions to disregard medical testimony about labial and vaginal penetration. One such explanation is the higher probability of gaining a conviction on lesser charges. In the period 1870–89, for example, acquittal rates for rape and carnal knowledge in the whole of England and Wales were an average of 56 per cent higher than those for indecent assault and assault with intent.[64] In 1888 *Lloyd's Weekly Newspaper* reported that, in a case of suspected carnal knowledge of an eight-year-old girl, the Wandsworth magistrate concluded that 'no jury would convict of that offence; but he should commit [the prisoner for trial] for an indecent assault upon this little girl'.[65] Carolyn Conley also found that some magistrates in Victorian Kent downgraded charges to try cases of alleged rape summarily, as they officially could only use summary justice in cases of indecent assault and common assault, a decision that also might also have related to conviction rates.[66] Another possible reason for pursuing lesser charges related to the 'respectability' of the accused in local contexts where reputation and character were crucial.[67] Although evidence of magistrates' decision-making processes is limited, it seems likely that a range of pragmatic concerns influenced their reception of medical testimony.

Unlike magistrates' courts, systematic manuscript evidence survives for grand juries' dismissal of cases. Grand juries of between 12 and 23 men had no ability to try a case but, like magistrates, could dismiss them as 'no bills' if they felt that there was not a *prima facie* (first appearance) case for trial. Officially the same property qualifications were required to serve as a grand juror and petty juror, but evidence from the regions under study indicates that David Bentley is right in noting that 'those selected were, by rateable value, of description of a better class than the ordinary common jurymen' and were certainly set apart from predominantly working-class complainants.[68] Unlike magistrates' courts, systematic manuscript evidence survives for cases that were dismissed by grand juries. Overall, the grand jury declared 'no bills' for 161 out of 1700 cases in Middlesex and 119 out of 928 cases from Gloucestershire, Somerset and Devon. Only 23 of these from Middlesex and 17 from the south-west counties had medical testimony, therefore the grand jury did not prevent a significant number of medical witnesses from reaching trial. Despite their limited number, 'no bills' provide further evidence of medical influence in pre-trial processes. The 'no bill' rate was significantly below average when medical witnesses found indicators of genital violence, which implies that their testimony played a role in the decisions to commit cases for trial: 'no bills' were found in 13 per cent of cases in which medical practitioners found no signs of genital violence, 7 per cent of cases in which medical practitioners declared uncertainty about interpreting bodily signs, and 2 per cent of cases in which medical practitioners found signs of genital violence.

The grand jury marked the final point at which a case could be dismissed before trial, after which it was passed forward to trial by petty jury (usually just described as a 'jury'). In his *Handbook of Forensic Medicine*, published in the 1860s, Johann Ludwig Casper noted that verdicts in criminal trials did not operate on the basis of 'strict proof' but rather on the 'mental conviction of the judge (or jury), attained by a consideration of all the ascertained facts in their entirety'.[69] These 'mental convictions' can only be understood by considering the social standing of members of the court. Exact records of the jurors involved in trials are rare but surviving Devon jury lists from 1901 and 1903 indicate that, supporting the findings of other historians, many petty jurors were of the lower-middle or middle classes.[70] Most decisions in the courts were therefore made by middle-class and upper-class males, of whom the judge was also one, who generally shared the gendered and class-based 'mental convictions' of medical practitioners.

Despite – or perhaps because of – overlaps between medical, legal and social concerns, it is extremely difficult to separate out the factors that shaped a verdict at trial.[71] Jurors might convict 'against the evidence', but also often acted under the direction of a judge.[72] Basic statistical analysis of correlations between petty jury verdicts and medical testimony reveals no noteworthy trends.[73] However, more significant patterns become evident when the age of a complainant is taken into consideration. Figure 2.1 shows that the correlation between medical testimony on genital violence and jury verdicts diminished gradually as complainants grew older. Rigid legal frameworks that treated childhood as a two-tier division, under the felony and misdemeanour clauses of sexual consent legislation, were followed as little by jurors as they were by medical practitioners. Although the sample size is too small to draw any certain conclusions, particularly for the youngest and oldest age groups, this age-based reception of medical testimony about genital violence was in line with generally higher conviction rates in cases involving younger girls and boys.[74] Medical evidence of genital violence increased the likelihood of a successful prosecution, as conviction rates in such cases were above average for all age groups, but there was no simple cause-effect relationship between medical evidence and trial outcomes. Jurors, like medical practitioners, demonstrated implicit mistrust of older complainants irrespective of the evidence provided at trial.

The connection between medical testimony on genital injury and trial verdicts was sometimes made extremely clear in newspaper reports.

Figure 2.1 Verdicts in cases with medical evidence of genital injury

Although newspapers cannot be read as direct - or even always accurate - accounts of court processes, they show that the presence and absence of injury was thought to be worth reporting in connection with verdicts. In 1870 *The Times* stated that in a Middlesex case of attempted carnal knowledge of a seven-year-old girl '[t]here was corroborative evidence by an inmate of the house and by the doctor, and the jury found a verdict of *Guilty*'.[75] In the pre-trial statement for this case, the surgeon had testified that 'the parts were considerably swollen as though violence had been used towards the child'.[76] Medical testimony on genital injury was similarly influential in cases involving boys, although they were much more rare. When a Middlesex surgeon found that a 15-year-old boy had 'considerable inflammation' and 'laceration' around the anus but that 'penetration had not been made', *The Times* reported that '[t]he evidence of the medical man strongly corroborated the story of the boy, and there was other confirmatory evidence in the case. The jury found the prisoner *Guilty*'.[77] In both of these cases the medical testimony of genital violence was cited as a direct influence on trial outcomes, albeit only alongside other corroborative evidence.

Medical testimony could also contribute to acquittals, as in a case from the Devon Quarter Sessions in which *Trewman's Exeter Flying Post* noted that '[t]he hearing of a charge of assaulting his daughter, aged 14, brought against Michael O'Brien, merchant sailor of Stonehouse, resulted in an acquittal, a doctor stating that there was no corroboration of the girl's story'.[78] *The Times* of 5 February 1852 reported on a Middlesex case in which 'the surgeon who examined [the complainant, aged 14] showed that the offence had not been attempted. The jury *acquitted* the prisoner'.[79] In both of these cases medical witnesses had testified to finding an absence of any genital injury. However, medical testimony never operated alone and in the latter case the acquittal was also because the girl's evidence 'differed from what she had previously given'.[80] Some juries used medical testimony about the absence of genital injury to guide their decisions about the final charge, as most courts offered a range of possible charges on which to convict. *The Times* reported that in another Middlesex case:

> Henry Margetson ... was indicted for having, on the night of Saturday, the 23rd of last month, assaulted with intent a girl of the age of 13 years ... The medical evidence did not support the more serious charge, and the jury found the prisoner *Guilty* on the count for indecent assault.[81]

Again, this report broadly corresponded to the pre-trial statement in which the surgeon had testified that 'I could find no marks of violence on the child at all'.[82] The jury used this medical testimony about an absence of marks of violence to downgrade the charge, but not to dismiss the case entirely as there was a witness to the girl screaming and running away from the prisoner.

All of these reports indicate a relatively unproblematic relationship between medical testimony and trial outcomes. However, the impact of medical testimony on jurors' decisions about charges was more complex than it first appears. When a Middlesex prisoner was accused of carnal knowledge of an 11-year-old girl in 1874, *The Times* noted that:

> [T]he evidence of the child and of the divisional surgeon of Police who had attended her, coupled with the prisoner's own statements, clearly established most abominable conduct on his part towards the child; but it did not appear that he had completed the offence. The jury found him *Guilty* of the attempt.[83]

Despite this claim that the prisoner had not 'completed the offence', the surgeon Thomas Jackman's pre-trial statement in this case included testimony that '[i]n my opinion penetration has to some extent taken place'.[84] Jackman's testimony indicated that the prisoner had 'completed the offence' in law by partially penetrating the complainant, even if full penetration had not occurred. This testimony corroborated the girl's claim that 'I am sure his person went into mine'.[85] There are three possible explanations for the discrepancy between medical testimony and verdict: the newspaper misreported the case; the judge did not direct the jury as to the fact that *partial* penetration constituted proof of 'carnal knowledge' by the court; or the jury chose wilfully to ignore medical testimony in the light of other evidence, such as the girl's failure to complain after the first offence and the prisoner's testimony that she was a consenting party.[86] The latter two possibilities indicate that judges or juries drew on wider frameworks of moral thought in order to interpret the medical testimony on genital violence that they received.

Medical evidence about genital violence also had some influence on judges' sentencing of prisoners. In the event of a prisoner being found guilty a chairman consulting with his 'brother justices' determined the sentence, as no trial went ahead without a minimum of two justices present.[87] Fewer people decided upon a sentence than a verdict, which meant that theoretically the sentencing process was more open to being swayed by the personal convictions of a court chairman and his colleagues. This theory is corroborated to an extent by the fact that the Middlesex Sessions' chairman from 1889 until 1908, Mr Littler, was apparently 'renowned in the contemporary press for his stiff sentencing practices'.[88] Andrew Ashworth notes that judicial discretion during sentencing was restricted in the early-nineteenth century because '[t]here were maximum and minimum sentences for many offences, and several statutes provided a multiplicity of different offences with different graded maxima'.[89] In the period and statutes under study, some of these restrictions were still in place. Few of the crimes had minimum sentences, unless a judge opted to impose penal servitude. However, after common law charges such as indecent assault and common assault were consolidated in the 1861 Offences against the Person Act, most sexual crimes were attached to a maximum sentence of two years imprisonment with or without hard labour. Any conviction on a reduced charge of common assault had a maximum sentence of one year.[90] Judges were therefore somewhat limited in the sentences that they could impose. Newspapers also often reported cases in which judges expressed a desire to sentence a prisoner to 'flogging' but were unable to do so.[91] However, judges still had great discretion within legal parameters and could allow prisoners to leave the court with little more than a fine. They also had the right to take into consideration character testimonials when making their decisions.

Judges regularly handed down higher sentences when a medical witness testified to finding that a young female victim had contracted venereal disease, which was perceived as contributing to the moral and physical 'ruin' of an innocent girl. As *The Lancet* observed in 1884, 'heavy sentences … are always inflicted when disease is communicated'.[92] After medical testimony that a ten-year-old girl had acquired syphilis from an assault, a Middlesex judge stated in 1867 that '[t]he pollution of a child of tender years was a crime the atrocity of which no language could describe'.[93] Although this judge referred to 'pollution' of both mind and body, the presence of venereal disease was part of his decision to hand out the maximum sentence of two years imprisonment with hard labour. The language of 'pollution' was linked to particular Victorian concerns about dirt and disorder, but concerns about the harmful nature of disease for a child were nothing new. Sarah Toulalan has found that early modern courts also handed out harsher sentences for cases involving the transmission of venereal disease.[94] In another late nineteenth-century case, in which the prisoner was also given the maximum sentence, a medical witness testified that a girl of only two years old was suffering from gonorrhoea and London's *Morning Post* noted the 'most distressing consequences' of the assault when reporting the sentence.[95] When four girls under ten years old were apparently assaulted in October 1851, the *Morning Post* reported that '[t]he worst feature in the case was, that up to the present time, the girls are suffering from the effects of the misconduct of the prisoner'.[96] This case involved evidence from a medical witness that the girls had gonorrhoea, which seems likely to have been the 'effects' to which the paper referred. However, in this case the judge considered the 'weak intellect' of the prisoner a mitigating factor and passed a sentence of nine months imprisonment. Medical testimony was always situated against other evidence, but venereal disease exacerbated a sexual crime in the eyes of both medicine and the law.

Medical evidence on genital injury demonstrates the range of moral and social concerns that shaped medical testimony on 'physical proofs'. In Middlesex and south-west England medical witnesses interpreted genital signs in the light of long-held ideas and existing expectations about the links between the body, age, race, class and gender. These medical practitioners integrated new medical research on the variable nature of developmental norms with older ideas and diagnostic techniques to reinforce the status quo, in which puberty was the life stage at which bodies naturally opened up and became sexual. Such testimony connected implicitly with general social concerns about child protection and with fears about the female body, with its emerging sexuality, at puberty. It fed into a medico-moral 'script' that spoke to the shared anxieties of magistrates, juries, judges and medical witnesses. Judges and juries responded positively to medical testimony on genital violence but also brought their own concerns to bear on trial outcomes. Many of these concerns were also moral in nature and overlapped with medical thought on the sexual body and the loss of innocence at puberty.

Case Analysis: Injury, 1861

The Case: This sample case explores, in greater depth, how the difficulties of interpreting genital discharge and diagnosing disease operated in the courts. Although an exceptional case in some ways, such as the number of medical witnesses and use of microscopic analysis, it also illustrates some of the general trends outlined in the accompanying chapter; the case speaks particularly to the theme of uncertainty and shows how age and class could interact in the diagnostic process. The case focuses on disease, but raises issues that have wide relevance to Victorian and Edwardian sexual forensics.

The Prisoner: John Martinitzi, aged 31, Engraver.
The Complainants: Elonia Thompson, aged eight, and Sarah Ann Thompson, aged 10.
The Complaint: Indecent assault (consent a defence) or attempted carnal knowledge (consent no defence).
The Pre-Trial Statement: Depositions taken 1 March 1861, Worship Street Police Court.[97]

Elonia Thompson: I live with my mother at 4 Whites Street, Bethnal Green Road. I am eight years old. I know the prisoner, he lodged in my mother's house. On Saturday night week he left. On Tuesday before that, in the evening after tea, he told me to carry up his candle into his bedroom. He was then downstairs in mother's parlour, my sisters were in there. I took it up. He went up with me into the back room on the second floor. He laid me on the bed. He undid his trousers and laid on me. Lifted up my clothes. I felt something like a finger in between my legs in front. He hurt me a little, I felt it inside me. He did not do it long. He took me down off the bed. He told me not to tell anyone to keep it to myself. I went downstairs. He came down and looked into my mother's room. I did not say anything to my mother till Sunday ... I went up into his room. He laid me on the bed ... On the Tuesday night I did not call out or cry at all. On Sunday I told mother. She took me to the Doctor. I was then taken to the station.

Sarah Ann Thompson: I live at 4 Whites Street, Bethnal Green Road. I know the Prisoner, he lodged in my mother's house ... On the Sunday evening I was in the Parlour, he put his hand up my clothes. My two sisters were in the room and my little brothers, he was sitting down, he sat me on his knee. He put his hand up my clothes then. On the next Wednesday in the afternoon, I took [a] frying pan up into his room. He told me to take it up. He came at the top of the stairs and called 'Sarah Ann, will you bring the frying pan up?' I took it up. He lifted me, laid me on the bed. He was standing at the side of it, he leaned upon me. I felt something under my clothes between my legs. It hurt me very much between the legs when he did it. He was not long doing it. I saw him buttoning his clothes up when he lifted me off the bed. I did not tell my mother that day. I did not call out while he was doing it. I told my mother on the following Sunday. On the Wednesday when I came down I saw something like matter on my clothes. I was then on the bottom stair.

Jane Thompson: I am a widow residing at 4 Whites Street, Bethnal Green Road. The witnesses Sarah Ann and Elonia Thompson are my daughters. On 28th last September Sarah was ten years old. On 27th last November Elonia was eight years old. I take in lodgers. The Prisoner lodges in the first floor back room that is the top room of the house. Last Saturday fortnight the 9th of February he came in. On the following Saturday in consequence of what had been said he left. On the Sunday following his leaving I noticed Sarah Ann's linen in a very dreadful state and her person as well ... Elonia's linen was marked and her person. I took her to the surgeon. I had examined her first and from what the surgeon said I examined Sarah Ann. I spoke to my daughters, they made a complaint to me. On the Monday following I took them to the Hospital. In consequence of what I heard, I gave the Prisoner in custody last night week. (cross-examined) I was out on Wednesday afternoon. I had not invited him into the Parlour on the Sunday previous. My son had.

Frederick Dawson: I am House Surgeon at the London Hospital. Sarah Ann and Elonia Thompson were brought in on the 18th of February. I examined them on that day. I examined Sarah [and] found inflammation of the private parts, no sign of laceration, no contusion whatever. No rupture of the hymen but a copious discharge which on examination proved to be pus. The clothes of the child were covered with discharge. A piece of her shift I took, it was covered with discharge. I examined the matter, found no evidence of spermatozoa but pus. Children are subject to such disease. Any disturbance of the digestive organs might cause such a disease or irritation of the parts. I think it possible that a person laying on the child and acting in the manner described might cause it. I examined Elonia likewise the same day I found her in a similar condition with the exception of not quite so much discharge. But in all other respects the same.

James Edmunds: Physician and Surgeon ... Residing at 2 Spital Square. On Thursday evening 21st about 9 in the evening. I examined the prisoner carefully and deliberately. I ascertained that he was suffering from no disease. I carefully inspected the urinary passage and ... found no evidence of present or past disease. I thought it possible he might recently have made water or wiped away the discharge. I caused him to be watched for two hours at the end of which time I examined him and found no reason to alter my former opinion. I examined the children the same evening. Sarah Ann, the elder, had a purulent discharge from the pudenda, but there was no evidence of laceration or contusion. She had a similar condition of the right eye, in technical language a catarrhal state showing that there was a tendency of catarrhal affection of the mucous membrane. With regard to the other child, there was no evidence of laceration or contusion and she was suffering in the same way, but more slightly. The discharge is probably from the natural causes. Children of weak nature are subject to that.

[Further examination. Taken 7th March 1861] Thomas Sarvis: I am a Doctor of Medicine residing at 1 Winchester Street, Waterloo Town, Bethnal Green. I know Sarah Ann and Elonia Thompson. On Sunday the 17th February last Elonia Thompson was brought to me. I examined her. Her pudenda was much inflamed and great deal of discharge was present of purulent character which I consider was gonorrhoeal. I subsequently on that day examined Sarah Ann Thompson. The pudenda was much inflamed and there was discharge of the same kind as her younger sister. I examined the linen of Sarah Ann Thompson. The discharge on it I believe to be gonorrhoeal discharge. I am aware that there is a disease from which children suffer of a similar nature but the discharge would not be so profuse I believe or of the same nature. I advised the mother to take the children to the hospital. Last Friday I saw the elder child she was then suffering from discharge. I examined her today and she is suffering from profuse discharge of the same kind. I examined the younger today and she is suffering from discharge less in quantity.

Ann Lane: I live at 4 Selvey Street East, Waterloo Town, Bethnal Green. I am the wife of John Lane a paper stainer. The prisoner formerly lodged with me five months in February. Four weeks ago next Saturday he occupied the back room first floor. I used to do all his washing. I noticed his shirts were dirty on the bottom part of the flap, messy white & yellow. I noticed them the whole time. The last shirt I did not notice. That was a week before he left – it was a coloured shirt – all the other shirts I washed for him were more or less marked. I did not notice any thing on the sheets. A short time after he came to live with me he told me he was going to the German Hospital. During that time I picked up two broken glass syringes in his room, in the fireplace. One I showed to my husband.

Police: I took the Prisoner in custody told him the charge. He said 'It was for nothing, it was out of spite'.

Prisoner: I have nothing to say.

The Verdict: Guilty of attempt to carnally know, 2 years hard labour (out of a maximum possible of 2 years).

Significance for Sexual Forensics: This case involved two young girls under the age of sexual consent and, in the absence of eyewitnesses, involved extensive attention to medical evidence. It was unusual in drawing upon the evidence of three medical witnesses. The mother of the complainants consulted a general practitioner and a hospital doctor, before the police called in their own divisional surgeon to examine the prisoner and re-examine the girls. These three witnesses were not, as in some cases, called 'for the prosecution' and 'for the defence' in order to be pitted against each other. All of the medical witnesses examined both complainants for signs of genital injury. The hospital surgeon also examined linen for signs of sperm, which was unusual in 1861 because it required the use of a microscope; based on all pre-trial statements, only 6 per cent of the provincial and 4 per cent of the Middlesex medical witnesses made any form of microscopic analysis in the entire period 1850–1914. Microscopical analysis involved the few medical practitioners who owned microscopes themselves or who had hospital appointments, as in this case, and generally focused on locating evidence of spermatozoa or semen. The medical witness in question found no such signs in his analysis, however, and the limited role of microscopic evidence at trial is indicated by the apparent importance placed instead upon lay witness testimony about 'messy white and yellow' marks on the prisoner's clothing.

As in most such cases, medical evidence focused primarily on the body. These three medical witnesses observed the same genital signs and acknowledged the same age-based issues, particularly those around common children's diseases, although did not come to the exact same conclusions. The hospital surgeon found no definitive signs of injury or disease when examining the girls. Despite finding a 'copious discharge', which might have been corroborative evidence of an offence, he situated this sign in relation to bodily norms and observed that 'children are subject to such disease'. The divisional surgeon of police more decisively supported the 'natural causes' explanation. He described carefully the process of proving the prisoner to be clear of disease and found both girls to be free of injury, except for a natural discharge. This practitioner also drew upon wider contemporary ideas about the links between age, class and disease. His testimony that such disease was common among '[c]hildren of weak nature' was indicative of a widespread belief, both in medicine and general society, that the working classes were more likely to suffer from natural discharges linked to uncleanliness and general weakness of constitution.

The local practitioner from Bethnal Green differed from the other two in testifying that the girls suffered from gonorrhoea. He acknowledged, under questioning, that 'I am aware that there is a disease from which children suffer of a similar nature' but stood by his diagnosis of venereal disease. That this practitioner gave a different diagnosis, the only one that provided definite corroborative evidence, shows the inherently ambiguous nature of sexual forensics when using naked-eye methods to interpret bodily signs. The diagnosis of gonorrhoea may have resulted from the medical practitioner's own clinical experience in such cases. It is also possible that medical witnesses were more likely to label a discharge as disease when combined with other markers of possible violence, here represented by an 'inflamed' external genital area (pudenda) that had apparently disappeared by the time of the divisional surgeon's examination. Contact with the girls' mother might have informed the medical diagnosis of gonorrhoea, as the mother testified that the discharge appeared only after the alleged assault. Her testimony indicated that this particular bodily sign was not the norm for the two girls in question, contradicting other medical witnesses who claimed that the complainants might be 'subject to' disease. The Bethnal Green practitioner also claimed to 'know' the girls, lived in their area and was the first point of contact for their mother; he may have been their family or local doctor with pre-existing knowledge of the girls' bodily norms. Whatever the explanation, his testimony shows that naked-eye diagnosis could lead to a range of explanations for the same genital sign, depending on issues such as time after the alleged assault and the individual doctor's experience.

The hospital surgeon's statement that '[a]ny disturbance of the digestive organs might cause such a disease or irritation of the parts. I think it possible that a person laying on the child and acting in the manner described might cause it' further highlights the complexities of medical testimony about genital signs. The medical witness gave broad support for two ostensibly contradictory theories. This approach did not represent ignorance, but rather acknowledged the ambiguous nature of the child's body in relation particularly to the distinctions between venereal disease, naturally-occurring disease and discharge caused by injury or inflammation. The hospital surgeon's willingness to accept a sexual crime as a possible explanation (without signs of extensive genital injury) may be explained by the nature of the alleged assault. Both girls' testimony indicated that the crime might have involved attempted penetration and/or manual penetration, not the more physically injurious act of sexual intercourse. As this medical witness did not testify to finding definite signs of penetration or venereal disease, his testimony could support a lesser charge of assault (in which the discharge was an indicator of injury rather than disease).[98] In 1861 the charge of 'indecent assault' still required evidence of resistance or dissent, however, meaning that only the charge of 'attempted carnal knowledge' could be pursued on this evidence.

Differences between medical testimony meant that the final verdict in this case was by no means inevitable; on the prisoner's side, newspapers implied that he was respectable and in court he blamed the charge on 'spite'.[99] However, the ambiguities of medical testimony about genital injury and discharge actually proved valuable, in allowing the jury to select the medical interpretation that best fitted with other evidence. The case had the support of the Associate Institution for the Protection of Women, there was consistency in the accounts of the two girls and the prisoner was accused of a very similar offence against another girl in Middlesex in the same Police Court session. In the light of this other evidence, medical evidence that supported the charge proved more convincing to the jury.

Notes

1. Francis Ogston, *Lectures on Medical Jurisprudence* (London: J. & A. Churchill, 1878), p. 124.
2. Only 25 medical witnesses from the sample commented on symptoms, such as complaints of 'pain', while all others focused on genital signs and 'Physical Proofs'.
3. For some discussions on the patient as 'object' and the rise of clinical observation, see Mary Wilson Carpenter, *Health, Medicine and Society in Victorian England* (Santa Barbara; Denver; Oxford: Praeger, 2010), p. 25; Anne Digby, *The Evolution of British General Practice, 1850–1948* (Oxford: Oxford University Press, 1999), p. 190; Mary E. Fissell, 'The Disappearance of the Patient's Narrative and the Invention of Hospital Medicine' in *British Medicine in an Age of Reform*, ed. Roger French and Andrew Wear (London; New York: Routledge, 1991), 92–109, p. 93.
4. Some of these methods were in development in the nineteenth century but were rarely implemented until police laboratories of the 1930s, the staff of which also educated local police forces in new forensic methods; see, for example, a Metropolitan Police Laboratory serologist's lecture to Exeter City Police; John C. Thomas, 'The Examination of Blood and Seminal Stains', *Police Journal* 10 (1937), 490–503.
5. Nick Lee, 'Faith in the Body? Childhood, Subjecthood and Sociological Enquiry' in *The Body, Childhood and Society*, ed. Alan Prout (Basingstoke: Palgrave Macmillan, 2000), 149–71, p. 154.
6. London, London Metropolitan Archives (LMA), Pre-Trial Statements, George Walker tried at the Middlesex Sessions on 5 December 1882 for indecent assault, MJ/SP/E/1882/050.
7. Alfred Swaine Taylor, *Medical Jurisprudence*, 4th edn (London: J. & A. Churchill, 1852 [1844]), p. 593. This quote was unchanged throughout all the editions, up to and including 1910.
8. Louise A. Jackson, 'Child Sexual Abuse and the Law: London 1870–1914', unpublished doctoral thesis, Roehampton Institute (1997), p. 180.

9. Taylor, *A Manual of Medical Jurisprudence*, 9th edn (London: J. & A. Churchill, 1874 [1844]), p. 678. This statement was a new addition to the 9th edition but was retained throughout all the others up to 1910.
10. Taylor, *Medical Jurisprudence*, 4th edn, p. 577.
11. Alfred Swaine Taylor, *The Principles and Practice of Medical Jurisprudence*, ed. Thomas Stevenson, 6th edn, vol. 2 (London: J. & A. Churchill, 1910 [1865]), p. 437.
12. London, LMA, Pre-Trial Statements, Charles Hutley tried at the Middlesex Sessions on 20 July 1869 for indecent assault, MJ/SP/E/1869/014.
13. London, LMA, Pre-Trial Statements, John Taylor tried at the Middlesex Sessions on 2 August 1881 for indecent assault, MJ/SP/E/1881/026.
14. Léon Henri Thoinot, *Medicolegal Aspects of Moral Offenses* (Philadelphia: F. A. Davis Company, 1911), p. 66.
15. Ogston, *Lectures on Medical Jurisprudence*, p. 105.
16. Gloucester, Gloucestershire Archives (GA), Pre-Trial Statements, Humphry Carpenter tried at the Gloucestershire Quarter Sessions on 1 July 1874 for attempted carnal knowledge, Q/SD/2/1874.
17. Gloucester, GA, Pre-Trial Statements, Nelson Daniel Short tried at the Gloucestershire Quarter Sessions on 23 August 1898 for indecent assault, Q/SD/2/1898.
18. 'Epitome of Current Medical Literature', *British Medical Journal (BMJ)*, 31 January 1903, 17–20, p. 19.
19. Helen King, *Hippocrates' Woman: Reading the Female Body in Ancient Greece* (London: Routledge, 1998), p. 71.
20. Ogston, *Lectures on Medical Jurisprudence*, p. 105.
21. Taunton, Somerset Heritage Centre (SHC), Pre-Trial Statements, Joseph Hilman tried at the Somerset Quarter Sessions on 4 July 1860 for carnal knowledge, Q/SR/639.
22. Exeter, Devon Record Office (DRO), Richard Wadge tried at the Devon Quarter Sessions on 1 July 1885 for indecent assault, Q/SD/2/1885.
23. London, LMA, Pre-Trial Statements, John Cleverley tried at the Middlesex Sessions on 19 February 1857 for indecent assault, MJ/SP/E/1857/003.
24. London, LMA, Pre-Trial Statements, Isaac Garcia tried at the Middlesex Sessions on 5 October 1883 for indecent assault, MJ/SP/E/1883/041.
25. London, LMA, Pre-Trial Statements, Garcia, MJ/SP/E/1883/041.
26. Exeter, DRO, Robert Hannaford tried at the Devon Quarter Sessions on 18 October 1876 for assault with intent, QS/B/1876/Michaelmas.
27. 'Devon Quarter Sessions', *Trewman's Exeter Flying Post*, 25 October 1876, 3, p. 3.
28. London, LMA, Pre-Trial Statements, Joseph Dungay tried at the Middlesex Sessions on 23 December 1869 for carnal knowledge, MJ/SP/E/1869/025.
29. London, LMA, Pre-Trial Statements, Dungay, MJ/SP/E/1869/025.
30. London, LMA, Pre-Trial Statements, Dungay, MJ/SP/E/1869/025.
31. London, LMA, Pre-Trial Statements, Dungay, MJ/SP/E/1869/025.
32. London, LMA, Pre-Trial Statements, Dungay, MJ/SP/E/1869/025.
33. William A. Guy and David Ferrier, *Principles of Forensic Medicine*, 5th edn (London: H. Renshaw, 1881 [1844]), p. 66. Only six per cent of medical witnesses in the South West and four per cent in Middlesex made any form of microscopic analysis.

34. Taunton, SHC, Pre-Trial Statements, John Henry Hayman tried at the Somerset Quarter Sessions on 1 July 1903 for indecent assault, Q/SR/812; see also London, LMA, Pre-Trial Statements, Dorothy Lund and Albert Warcup tried at the Middlesex Sessions on 22 June 1886 for indecent assault, MJ/SPE/1886/031.
35. Simmonds (1914) 9 Cr App R 51.
36. Robert Gray, 'Medical Men, Industrial Labour and the State in Britain, 1830–50', *Social History* 16 (1991), 19–43, p. 38.
37. 'The Effeminacy of the Latin Race', *The Lancet*, 12 February 1898, 447–48, p. 447.
38. John Roberton, 'On the Alleged Influence of Climate on Female Puberty in Greece', *Edinburgh Medical and Surgical Journal* 62 (1844), 1–22.
39. Sarah Toulalan, 'Introduction' in *Bodies, Sex and Desire from the Renaissance to the Present*, ed. Kate Fisher and Sarah Toulalan (Basingstoke: Palgrave Macmillan, 2000), 1–26, p. 15; Sara Read, Menstruation and the Female Body in Early Modern England (Basingstoke: Palgrave, 2013), p. 49.
40. George R. Drysdale, *The Elements of Social Science; or Physical, Sexual and Natural Religion. An Exposition of the True Cause and Only Cure of the Three Primary Social Evils: Poverty, Prostitution, and Celibacy*, 25th edn (London: E. Truelove, 1886 [1854]), p. 66.
41. London, LMA, Pre-Trial Statements, Robert Brown tried at the Middlesex Sessions on 19 January 1875 for indecent assault, MJ/SP/E/1875/002; Taunton, SHC, Pre-Trial Statements, Abraham Escott tried at the Somerset Quarter Sessions on 28 June 1871 for indecent assault, Q/SR/684.
42. On the absence of a 'bacteriological revolution' more generally see Michael Worboys, 'Unsexing Gonorrhoea: Bacteriologists, Gynaecologists and Suffragists in Britain, 1860–1920', *Social History of Medicine* 17 (2004), 41–59; Michael Worboys, 'Was there a Bacteriological Revolution in Late Nineteenth-Century Medicine?', *Studies in History and Philosophy of Biology and Biomedical Sciences* 38 (2007), 20–42.
43. Taylor, *Principles and Practice*, ed. Frederick J. Smith, 5th edn, vol. 2 (London: J. & A. Churchill, 1905 [1865]), p. 135.
44. See 'The Micrococcus of Gonorrhoea', *The Lancet*, 22 November 1884, 927, p. 927; 'The Practical Value of the Gonococcus', *The Lancet*, 16 April 1887, 790, p. 790. These concerns fed directly into the 1891 edition of Taylor's forensic medicine text, in which he referred to doubts about 'the specific character of the micrococcus'; Taylor, *Manual*, 12th edn (London: J & A Churchill, 1891 [1844]), p. 701.
45. For example in F. Swinford Edwards, 'The Treatment of Gonorrhoea with Special Reference to Bladder Irrigation', *The Lancet*, 12 April 1902, 1029–31, p. 1029.
46. Exeter, DRO, John Clements tried at the Devon Quarter Sessions on 3 July 1884 for indecent assault, QS/B/1884/Midsummer; London, LMA, Pre-Trial Statements, William Hunter tried at the Middlesex Sessions on 21 August 1884 for indecent assault, MJ/SP/E/1884/038.
47. Ogston, *Lectures on Medical Jurisprudence*, pp. 94–95.
48. London, LMA, Pre-Trial Statements, Alfred Boydry tried at the Middlesex Sessions on 20 August 1877 for indecent assault, MJ/SP/E/1877/019.

49. London, LMA, Pre-Trial Statements, Allwin Musgrove tried at the Middlesex Sessions on 2 May 1882 for indecent assault, MJ/SP/E/1882/015.
50. *The Rational Key to Health & Long Life: Comprising a Concise Treatise of Physiology and a Plain Exposition of the Causes, Symptoms and Treatment of Diseases with Remarks on the Antibilious and Curative Power of Paul Gage's Elixir*, 2nd edn (Paris: Paul Gage, c. 1859).
51. London, LMA, Pre-Trial Statements, Edwin Kew tried at the Middlesex Sessions on 28 February 1882 for indecent assault, MJ/SP/E/1882/008.
52. For example London, LMA, Pre-Trial Statements, William Camp tried at the Middlesex Sessions on 24 September 1860 for indecent assault, MJ/SP/E/1860/019.
53. 'Vulvar Discharge in Children and Charges of Rape', *BMJ*, 21 January 1888, 162, p.162.
54. A high-profile study by Dr W. B. Ryan apparently confirmed the sponge as a possible mode of transmission; see Taylor, *Medical Jurisprudence*, pp. 580–81.
55. London, LMA, Pre-Trial Statements, James Westhall tried at the Middlesex Sessions on 29 September 1868 for carnal knowledge, MJ/SP/E/1868/017.
56. *Seduction Laws Amendment Bill: Report of the Debate on this Bill in the House of Commons* (Manchester: Alexander Ireland and co., printers, 1873), p. 7.
57. W. T. Stead, 'The Maiden Tribute of Modern Babylon: the Report of our Secret Commission', *Pall Mall Gazette*, 6–10 July 1885.
58. Gayle Davis, *'The Cruel Madness of Love': Sex, Syphilis and Psychiatry in Scotland, 1880–1930* (Amsterdam: Rodopi, 2008), p. 241.
59. Roger Davidson, ' "This Pernicious Delusion": Law, Medicine, and Child Sexual Abuse in Early-Twentieth-Century Scotland', *Journal of the History of Sexuality* 10 (2001), 62–77, pp. 68–69.
60. Davidson, ' "This Pernicious Delusion" '.
61. As noted in the book's introduction, the felony age category applied to girls under ten until 1875, under 12 until 1885 and from then onwards to girls under 13 years old; London, LMA, Pre-Trial Statements, Alfred Robinson tried at the Middlesex Sessions on 24 August 1874 for indecent assault, MJ/SP/E/1874/017; Exeter, DRO, John Fisher tried at the Devon Quarter Sessions on 29 June 1871 for indecent assault, QS/B/1871/Midsummer.
62. Louise Jackson, *Child Sexual Abuse in Victorian England* (London: Routledge, 2000), pp. 78–79.
63. John Adams, 'What Acts are Essential to Constitute Rape?', *The Lancet*, 25 March 1843, 933, p. 933.
64. British Parliamentary Papers (BPP), Judicial Statistics of England and Wales for 1870–1889 (London: H.M.S.O., 1871–1890).
65. 'Police Intelligence', *Lloyd's Weekly Newspaper*, 22 April 1888, 4, p. 4.
66. Carolyn A. Conley, 'Rape and Justice in Victorian England', *Victorian Studies* 29 (1986), 519–36, p. 521.
67. Carolyn A. Conley, *The Unwritten Law: Criminal Justice in Victorian Kent* (Oxford: Oxford University Press, 1991), p. 83; Kim Stevenson, '"Unequivocal Victims": The Historical Roots of the Mystification of the Female Complainant in Rape Cases', *Feminist Legal Studies* 8 (2000), 343–66.
68. David Bentley, *English Criminal Justice in the Nineteenth Century* (London: Hambledon Press, 1998), p. 148; Exeter, DRO, Devon Epiphany Sessions: The Names of the Grand Jurors, 1901, QS/B/1901/Epiphany; Exeter, DRO, Devon

Michaelmas Sessions: The Names of the Grand Jurors, 1903, QS/B/1903/Michaelmas.
69. Johann Ludwig Casper, *Handbook of Forensic Medicine*, trans. from 3rd edn by G. W. Balfour, vol. 1 (London: New Sydenham Society, 1861), p. vi.
70. See the discussion of middle-class jurors in Martin J. Wiener, *Men of Blood: Violence, Manliness, and Criminal Justice in Victorian England* (Cambridge: Cambridge University Press, 2004), p. 38.
71. Some influential factors, such as witness conduct and the direction of judges, are not recorded in depositions. Newspapers fill some of these gaps, but are far from a direct lens into jurors' decision-making processes. They tell us as much about the construction of stereotypes as 'real' events in the courtroom; see Kim Stevenson, 'Unearthing the Realities of Rape: Utilising Victorian Newspaper Reportage to Fill in the Contextual Gaps', *Liverpool Law Review* 28 (2007), 405–23.
72. Although few records survive of judges' final comments or directions, contemporary news reports indicate that 'directed verdicts' were widespread. For example of directed verdicts reported in newspapers see 'Somerset Quarter Sessions', *Bristol Daily Mercury*, 11 April 1907, 6, p. 6; Devon Midsummer Sessions', *Trewman's Exeter Flying Post*, 8 July 1868, 3, p. 3; 'Middlesex Sessions', *Lloyd's Weekly Newspaper*, 29 August 1886, 4, p. 4.
73. As the average conviction rate for the two regions under study was 67 per cent, it is ostensibly significant that prisoners were convicted in 78 per cent of cases in which medical practitioners found marks of genital violence on females or males. However, prisoners were also convicted in 68 per cent of cases in which medical testimony on genital signs was uncertain, and 72 per cent of cases in which *no* marks of genital violence were found.
74. Overall, for the same five age groups depicted in Figure 2.1 from youngest to oldest, the average conviction rates were 79 per cent, 69 per cent, 73 per cent, 64 per cent and 63 per cent respectively. These statistics are calculated from 'true bills' without guilty pleas in which a complainant's age was known, as follows: 0<4 years old: 11 guilty, 3 not guilty; 4<8 years old: 148 guilty, 64 not guilty, 1 insane; 8<12 years old: 409 guilty, 151 not guilty, 1 insane; 12<16 years old: 199 guilty, 110 not guilty, 1 insane; 16+ years old: 90 guilty, 51 not guilty, 1 insane.
75. 'Middlesex Sessions', *The Times*, 15 March 1870, 11, p. 11.
76. London, LMA, Pre-Trial Statements, Thomas Austin tried at the Middlesex Sessions on 14 March 1870 for attempted carnal knowledge, MJ/SP/E/1870/005.
77. London, LMA, Pre-Trial Statements, Justus Dickhart tried at the Middlesex Sessions on 7 June 1864 for indecent assault on a male, MJ/SP/E/1864/011; 'Middlesex Sessions', *The Times*, 8 June 1864, 13, p. 13.
78. 'Devon Quarter Sessions', *Trewman's Exeter Flying Post*, 21 October 1905, 7, p. 7.
79. 'Middlesex Sessions', *The Times*, 5 February 1852, 7, p. 7.
80. 'Middlesex Sessions', *The Times*, 5 February 1852, 7, p. 7.
81. 'Middlesex Sessions', *The Times*, 16 November 1875, 9, p. 9.
82. London, LMA, Pre-Trial Statements, Henry Margetson tried at the Middlesex Sessions on 15 November 1875 for indecent assault, MJ/SP/E/1875/022.
83. 'Middlesex Sessions', *The Times*, 21 October 1874, 11, p. 11.

84. London, LMA, Pre-Trial Statements, William Paget tried at the Middlesex Sessions on 20 October 1874 for carnal knowledge, MJ/SP/E/1874/022.
85. London, LMA, Pre-Trial Statements, Paget, MJ/SP/E/1874/022.
86. This list includes the possibility that a judge may have *omitted* certain information when directing a jury but does not include the possibility of deliberate misdirection, because the latter would likely have been noted in newspapers, medical literature and/or an appeal.
87. Charles Knight, 'Quarter Sessions', *The Penny Magazine of the Society for the Diffusion of Useful Knowledge*, 31 January 1839, 40, p. 40.
88. C. E. A. Bedwell, 'Littler, Sir Ralph Daniel Makinson (1835–1908)', rev. Eric Metcalfe, *Oxford Dictionary of National Biography* (Oxford: Oxford University Press, 2004) <http://www.oxforddnb.com/view/article/34557> (accessed 23 July 2015).
89. Andrew Ashworth, *Sentencing and Criminal Justice*, 5th edn (Cambridge; New York: Cambridge University Press, 2010 [1992]), p. 52.
90. Young offender legislation also allowed for prisoners under the age of 16 to be sent to reformatory school for between two and five years, although this power was rarely used in the cases under study.
91. 'Law and Police', *Reynolds's Newspaper*, 18 September 1870, 8, p. 8.
92. 'Liverpool (From Our Own Correspondent)', *The Lancet*, 24 May 1884, 963, p. 963.
93. London, LMA, Pre-Trial Statements, Robert White tried at the Middlesex Sessions on 22 January 1867 for carnal knowledge, MJ/SP/E/1867/002; 'Middlesex Sessions', *The Times*, 23 January 1867, 9, p. 9.
94. Sarah Toulalan, ' "Is He a Licentious Lewd Sort of a Person?": Constructing the Child Rapist in Early Modern England', *Journal of the History of Sexuality* 23 (2014), 21–52, p. 38.
95. London, LMA, Pre-Trial Statements, George James tried at the Middlesex Sessions on 17 February 1858 for indecent assault, MJ/SP/E/1858/004; 'Middlesex Sessions', *Morning Post*, 18 February 1858, 7, p. 7.
96. London, LMA, Pre-Trial Statements, Joseph Willis Ford tried at the Middlesex Sessions on 20 October 1851 for assault with intent, MJ/SP/E/1851/022; 'Middlesex Sessions', *Morning Post*, 22 October 1851, 7, p. 7.
97. London, London Metropolitan Archives (LMA), Pre-Trial Statements, John Martinitzi tried at the Middlesex Sessions on 22 March 1861 for attempted carnal knowledge, MJ/SP/E/1861/006. Full transcript, except for occasional sentences with illegible words. Some punctuation added for clarity.
98. Although this practitioner did not explicitly reject venereal disease as an explanation, his complete lack of reference to the topic indicates a greater support for the 'irritation of parts' theory (whether from natural causes or injury) than venereal disease.
99. 'Police', *The Times*, 15 March 1861, 11, p. 11.

3
Innocence: Chastity and Character

The violated body was a medical, legal and social text: medical men needed to read the body and its signs in order to give them meaning. As part of this process, medical witnesses considered and eliminated a range of different explanations for physical signs and symptoms. One such explanation was the complainant's sexual history, a subject that was considered to be irrelevant in law but was a necessary part of the diagnostic process. As with bodily signs in general, medical testimony about chastity and sexual character differed according to the age of complainants, but showed no clear division between 'children' and 'adults'. Medical witnesses focused on the subjects of masturbation and genital 'play' in relation to the youngest complainants, and only contemplated unchastity as an explanation for genital signs once girls neared or passed the age of puberty. Examining the age, class and gender dimensions of medical testimony on masturbation and unchastity – and courts' reactions to such evidence – shows that forensics provided a scientific justification for dismissing cases that did not fit age-based and gendered stereotypes of innocent victimhood.

For girls of different ages, medical witnesses gave alternative explanations for the same bodily sign associated with chastity: the hymen. The intact hymen was a crucial indicator of virginity and symbol of the enclosed, untouched body. Although medical literature emphasised the need for general caution when interpreting such signs, the forensics expert Johann Ludwig Casper emphasised that 'we must not be thereby led astray in determining the value of this sign, which is *the most valuable of all in a diagnostic point of view*'.[1] Like many writers on medical jurisprudence, Casper went on to note that the absence of a hymen was a likely mark of genital violence if so-called *carunculae myrtiformes* were present. These small fleshy remnants of a perforated hymen apparently shrivelled over time, therefore provided evidence not only of a sexual encounter but also of when the encounter occurred. The absence of a hymen without *carunculae myrtiformes* was a sign of possible previous unchastity or of repeated sexual assaults over a long duration, as often claimed in incest cases. Medical practitioners claimed, building upon early modern humoral theory, that such repeated assaults could open the vagina sufficiently to allow for sexual pleasure; such a victim was thought to lose the bodily characteristics of childhood and become a form of social misfit.[2]

There was clear medico-legal advice on the value of the hymen as a sign, but medical witnesses interpreted the same signs differently according to the age of a complainant. The younger the girl, the less likely medical witnesses were to attribute an absent hymen without *carunculae myrtiformes* to unchastity or long-term abuse, instead preferring natural absence as an explanation. No case reached trial in which a medical witness found a girl below the age of ten to have been previously unchaste. The absence of such cases from trial can be interpreted in two ways: either medical witnesses did not consider unchastity to be a possibility among such young girls, or these cases were all dismissed at magistrates' courts. The first explanation seems more compelling because medical witnesses at trial often found that young girls had damaged or absent hymens, but failed even to acknowledge sexual intercourse as a possible explanation. The complete absence of these cases also supports this interpretation; if medical testimony about the unchastity of very young girls existed, at least some of the cases should have reached trial on the basis of other witness testimony.

Despite finding an absent or imperfect hymen in girls as young as three, without the presence of *carunculae myrtiformes*, medical practitioners attributed this sign to natural causes or 'congenital' absence in cases with the youngest complainants. In a Middlesex case from 1876, for example, a medical witness examined a 'child' of unspecified age and testified that 'I believe there never had been a hymen – it was congenitaly [sic] deficient'.[3] Medical witnesses also repeatedly emphasised the 'small' nature of hymens in young girls. The small hymen represented a child's generally asexual body, which was closed off by the hymen but already so small as to be almost impenetrable. In 1877, when a seven-year-old girl was allegedly raped by a lodger, the divisional surgeon of police testified that 'the hymen is not I think ruptured ... there was a small hymen' and a second medical witness testified that '[t]he hymen which was very small & inconspicuous was in my opinion intact'.[4] For girls so young, it was not unusual to have a hymen so small as to be almost undetectable; in symbolic terms, a small hymen both served to protect a child's innocence and to represent the limited extent of their sexuality. The possibility of very young girls consenting to having sexual partners was thus removed, at least in the majority of cases. Although medical practitioners never took this line, some prisoners did blame young children for initiating sexual contact. In an exceptional case from Gloucestershire, a prisoner described a six-year-old girl as a 'little whore' and a defence witness describe the same child as 'very forwards'.[5] However, a newspaper described the prisoner's claims as 'extraordinary allegations' and the medical witness testified to finding 'nothing about [the] child's parts to suggest she was a whore to a professional man'.[6]

Medical witnesses articulated few doubts about the innocence of the very young, but outside of the courts medical practitioners demonstrated great concern about the implications of their 'fall'. 'Fallen' girls were thought likely to become sexually precocious and, in the worst-case scenario, to fall into prostitution.[7] The precocity or 'pollution' of young girls, even those who were victims of crime, also then posed an apparent threat to those around them. As popular medical writer William Acton complained, 'systematic seducers ... pollute the mind of modest girls'.[8] This language of 'pollution' articulated particular concerns of late-Victorian and Edwardian society. It symbolised dirt, disorder, immorality and urban 'pathology' that the middle classes increasingly feared for a range of political, social and economic reasons.[9] In Mary Douglas's anthropological work on pollution beliefs, she noted that the rhetoric of dirt and uncleanness is based on the principle of 'matter out of place'.[10] Sexual contact with a young girl was 'out of place' both because it was not within a marital relationship and because it predated the girl's maturity. These concerns about the 'fallen' child did not serve to sexualise the 'normal' child, but rather to emphasise the undesirable moral consequences of sexual crime. Medical men and high-profile social purity, feminist and child protection groups used the pathological 'abnormal' sexual child to emphasise the inherent innocence of the majority.

The discussions around 'fallen' girls and precocity indicated that children would only be sexually active as a result of a previous sexual assault, rather than beforehand. As it was relatively rare for the same complainant to come before the courts twice, in which circumstance she would be seen as a victim in the first case and potentially precocious in the second, medical witnesses generally rejected sexual intercourse as an explanation for the absence of a hymen in the young.[11] However, innocence and sexuality were part of a spectrum rather than a binary. Medical practitioners assumed that most young girls and boys would be virgins, but did sometimes raise the question of masturbation in such cases.

In 1887 a Gloucestershire surgeon found a six-year-old girl's hymen to be ruptured, but admitted under cross-examination that 'I have known cases of children playing with themselves and thus being ruptured'.[12] This testimony represented a genuine belief that young girls might 'play' with their genitals, although the medical witness thought it an unlikely explanation for this particular complainant's bodily signs. He declared his opinion that the hymen was ruptured from violence and refuted a suggestion, put forward by the prisoner's defence, that it might have been damaged 'with a rough stone' from falling down. The medical witness evidently was not willing to accept infinite explanations for a rupture of the hymen, but acknowledged genital 'play' as a possibility. The testimony acknowledged the ambiguities of the hymen as a sign, of sexual development and of sexual behaviour. It related to contemporary anxieties about masturbation, but without explicit terms such as 'self-abuse'. In discussing genital 'play' the medical witness recognised that a young girl might engage in ostensibly sexual acts without sexual intent. The language of 'playing with' herself was desexualised and innocent; it represented a general belief that young children might explore their bodies without being aware of the implications of their actions. Other medical witnesses used similar language, as in a Middlesex case from 1877 in which – in relation to a case of alleged assault on a girl aged seven – the medical witness testified under cross-examination that 'I know as a medical reader that children do play with their private parts'.[13]

Medical witnesses were broadly open to the possibility that young girls and boys might self-induce signs of injury through masturbation, but did not always find evidence of such habits. In one case from Somerset in 1903 a general practitioner refuted the defence counsel's suggestion that an eight-year-old girl could have caused her own genital injuries, in this case tenderness and bloodstains, by stating that:

> The bruise was at the place of excitation on the private parts ... There was no enlargement of the organ, which would have appeared had the child been in the habit of using excitement. A habit of excitement by a child would show general enlargement.[14]

Like the Gloucester case, the prisoner's defence posed questions about the 'habit of excitement' alongside other possible explanations, including early menarche and 'rocking in a rocking chair'. Again, the medical witness attributed the signs to violence and denied that the rocking action would cause bleeding, but showed a broad willingness to entertain the notion of a 'habit of excitement' in an eight-year-old girl. Such language indicates that young children could retain their status as innocent victims despite engaging in masturbatory acts. In medical, social and legal language they were not engaging in conscious 'self-abuse' but instead in a more physical 'play' or 'excitement'. These phrases connected with contemporary literature on child development, which denied a self-aware sexuality in the young but acknowledged children's growing potential for sexual sensations.

Medical writers evaded making links between childhood masturbation and any innate form of 'sexuality'. They often emphasised that young boys and girls masturbated for reasons such as illness, immoral influences or precocious sexual instincts that came significantly before the ability to understand it.[15] Elizabeth Blackwell's *The Human Element in Sex*, for example, rejected the possibility that children were responsible for the 'sensations' that led to masturbation, or that they were aware of the implications of their actions:

> [I]n the little ignorant child this habit springs from a nervous sensation yielded to because, as it says, 'it feels nice.' The portion of the brain which takes cognizance of these sensations has been excited, and the child, in innocent absence of impure thought, yields to the mental suggestion supplied from the physical organs. This mental suggestion may be produced by the irritation of worms, by some local eruption, by the wickedness of the nurse, occasionally by malformation or unnatural development of the part themselves.[16]

The 'mental suggestion' to which Blackwell referred was less an active awareness than a decision made on the basis of bodily sensations, which developed in consequence of pathological symptoms or 'evil example'. Concerns about negative environmental influences rather than innate sexual urges were central to her work, and the work of many of her contemporaries. Important publications such as *The Lancet* similarly described girls as young as five as 'victims' of masturbation rather than as agents of a sexual act.[17]

Only a very small group of medical authors suggested that an inclination to masturbate could be innate in some girls and boys. In William Acton's 1857 work on *The Functions and Disorders of the Reproductive Organs* ..., which was extremely successful with the lay public despite being originally intended for medical colleagues, he stated that:

> [L]ittle doubt exists in my own mind, that in some precocious children sexual ideas may become developed many years previously to the perfect evolution of the genital organs ... It has been supposed that this depends upon improper excitement of the sexual organs by nursemaids. That such may often be the case, I can quite believe; but I feel certain that very young children may inherit a disposition to affections of these organs, which causes them to rub themselves and incidentally to excite abnormal sensations and partial erections.[18]

Acton did not deny that precocious sexual ideas could be the result of children being led astray by nurses or of pathological conditions that drew attention to the genitals; his use of 'precocious' indicated that he thought these children to be abnormal. However, Acton also promoted the idea that precocity could be the result of 'heritable' conditions and therefore be normal for certain types (or class) of child.

Followers of Freud's work on sexuality, translated into English in the early-twentieth century, believed even more strongly that the infantile 'sexual instinct' was innate. However, Steven Marcus rightly notes that 'of all Freud's findings those that have to do with infantile and childhood sexuality were resisted with the most persistency'.[19] Despite the popularisation of Freud's theories, in 1913 a review in the *British Medical Journal* complained that 'some of [Freud's] conclusions appear somewhat grotesque. Freud believes and teaches that sexual sensations become obtrusive at a very early age – in very earliest infancy'.[20] Freud's ideas about innate infantile sexual 'instincts' had an ambivalent reception and only a slow acceptance in mainstream medicine. They certainly had no impact outside scientific communities before the First World War and seemingly had no impact at all on sexual forensics.

In general, medical witnesses in court aligned more with Blackwell than with Acton or Freud throughout the period. They acknowledged the possibility of self-play among children, but desexualised such actions and did not conceptualise them as 'self-abuse'. Although not unusual, the habit was certainly not thought to be typical. Medical witnesses also did not link a habit of 'excitement' or 'play' to heredity. The difference between a habit grounded in physical pleasure and an inherited habit was significant in relation to ideas about victimhood: in the first framework of thought, a child was innately innocent but curious, whereas in the second framework of thought they were innately sexual but potentially unaware. Only the first approach allowed medical witnesses to acknowledge child 'play' as a possible cause of genital injury without destabilising the child's position as innocent 'victim'.

The innocent language of 'play' evolved into a more neutral tone for older children, including those under the age of sexual consent. In a case of alleged indecent assault from Middlesex in 1865, in relation to a nine-year-old female complainant, a police surgeon testified that:

> I feel confident that the laceration of the hymen must have been occasioned by the introduction of some foreign body: it might have been occasioned by the introduction of a man's finger ... ([cross-examined] by Mr Young) Coughing would not cause the rupture – it might have been done by a child's finger.[21]

The surgeon's testimony, under cross-examination, that the girl's lacerated hymen 'might have been done by a child's finger' raised questions about the value of his corroborative evidence. This line of defence questioning implied that she may have inserted a finger into her vagina, at least sufficiently far to rupture the hymen, although without considering whether such an action would be out of 'playful' curiosity or indecent intent. As girls grew older, edging closer to the age of puberty, such actions were no longer necessarily innocent 'play' but did not fall automatically into the category of aware 'self-abuse'.

In another case involving a girl aged ten, tried in Middlesex in 1868, under cross-examination the medical witness acknowledged the possibility that the absence of a hymen resulted from the complainant's 'own act'. This phrase implied greater agency than 'play' or 'excitement' and was reflective of a belief that girls gradually lost their innocence with age. This medical witness also noted the increased ambiguity of the hymen as a bodily sign as complainants grew older, with a growing number of possible explanations for its absence:

> [T]here was no hymen, but it had not been recently broken. It is an unusual thing to have no hymen at that age, it may be absent from natural causes. (cross-examined by Mr Pelham) ... I cannot say how long the hymen had been wanting. The redness of the parts might have been caused of her own act, but not by walking. I don't know how long this may have existed.

This medical testimony, that an absent hymen was 'unusual' at the age of ten, implied that girls were unlikely to be sexually active before puberty. However, the hymen was an ambiguous sign and its absence could also be attributed to 'natural causes' or to 'her own act'. In this case, as in many others, the prosecution and defence took advantage of the ambiguities of the hymen as a sign and elicited explanations that supported their respective cases. The shift from 'play' to 'act' did not only represent changes in medical thought, but was also a defence strategy. Prisoners' representatives in court subtly encouraged medical witnesses to describe girls as actively sexual or precocious, rather than innocently exploring, particularly as complainants neared the expected age of puberty.

The hymen as a sign became increasingly ambiguous as its possible meanings multiplied, which happened gradually as a girl grew older. For a complainant aged ten, the medical witness acknowledged a range of possible explanations for an absent hymen beyond asexual 'play' but did not yet explicitly include unchastity as an explanatory factor. A year or two closer to puberty, however, medical writers and witnesses began to include unchastity as a possibility and spoke more explicitly of 'masturbation' for girls and 'self-abuse' for boys. Within medical jurisprudence literature of the late-nineteenth century, writers such as Francis Ogston noted that:

> [I]t is also necessary to remind you of some of the changes which may be effected on the state of the genitals, which in the virgin adult female may simulate certain of the consequences of sexual connection. I refer to the effects which may be produced by the practice of the solitary vice of masturbation, but no means a rare one with respectable females.[22]

These comments connected with wider studies of sexual development, which explained the apparent prevalence of masturbation at and after puberty. Physiologists and alienists alike emphasised that the physical ability to experience sexual pleasure could develop long before psychological maturity.[23] Pubescent girls and boys were therefore thought to be at particular risk of falling into immoral habits as, despite their natural capacity to experience pleasure, they lacked a complete understanding of sex or the respective 'modesty' and 'will' necessary for self-restraint.[24]

General social anxieties about masturbation had long focused on boys and medical witnesses articulated particular concerns about boys approaching the age of expected puberty. In one Middlesex case involving a 13-year-old male complainant, tried in 1875, two medical practitioners testified under cross-examination that a part of the boy's penis was lacerated but that 'I do not think the appearance I saw could have been caused by the boy himself' and 'I do not think the state of the boy's penis could have been produced by self-abuse. It might have been'.[25] Here, the language of medical testimony and cross-examination shifted again. A 13-year-old boy engaging in 'self-abuse' was at fault and knowing of his actions; the language of 'self-abuse', 'excitement' and 'playing' was distinct and appropriate for each age group. The term 'abuse' related to the growing likelihood, and perceived implications, of harm at the age of puberty. At this age, boys apparently had not yet developed self-control but might have semen to 'waste', therefore were more susceptible to diseases like spermatorrhoea.[26] In 1890 *The Lancet* also highlighted the particular dangers to older boys when it stated that '[i]n boys, especially those nearing puberty, there is considerable evidence to show that masturbation may not only provoke paroxysms of palpitation, but may lead to conditions of irritable heart'.[27]

In medical thought masturbation was a natural consequence of sexual development, because sexual curiosity emerged at a time of limited willpower, but there was also a strong social and medico-moral tradition of describing masturbation as 'unnatural'. Girls and boys apparently could resist temptation with the correct moral guidance, therefore masturbation was associated with ignorance, immoral surroundings and the lower classes. Such class prejudices had long roots, stemming back to eighteenth-century works on onanism that highlighted the extensive nature of the habit among servant girls.[28] Medical and social writers alike couched masturbation in terms of the unnatural 'wasting' or 'spilling' of seed. These ideas represented religious concerns, the continuation of long-term medical concerns about the health risks of wasting spermatozoa and humoral theories about the importance of balancing bodily fluids.[29] The influence of medical literature on 'self-abuse' was not linked only to moral matters, but also to the culturally-specific relevance of 'waste' in the Victorian and Edwardian periods. As Sally Shuttleworth notes, '[i]n an industrial culture governed by moral, economic, and psychological ideologies of self-control and the efficient channelling of energy, masturbation, that wasteful and hidden practice, came to seem the ultimate sin of childhood'.[30] With revelations about the physical fitness of the race, most famously the general poor health of recruits in the Boer War, masturbation came to symbolise wider concerns about masculine weakness. The emergence of eugenics, an increasingly popular multidisciplinary field of enquiry in the early-twentieth century, further fuelled concerns about the implications of 'wasted' energy for the future of the nation.

As one teacher wrote in *Eugenics Review* in 1913, a boy should be told that his new sexual and bodily force 'should not be wasted ... It affects not only the individual, but the community and the race'.[31]

When faced with older complainants, medical practitioners not only expressed more concern about masturbation as a form of active sexual behaviour, but also focused increasingly on unchastity. Although unchastity had long been frowned upon for religious reasons, it became a social issue of particular note in the late-nineteenth and early-twentieth centuries. In addition to the lingering influence of Puritanism and Evangelicalism, demographic trends drove concerns about pre-marital sex and illegitimacy. The proportion of unmarried women rose from around ten per cent in the mid-nineteenth century to 14 per cent in the first third of the twentieth century, while fertility fell consistently over the same period.[32] Notions that the working classes were breeding, while the middle classes were shrinking, fuelled concerns about the decline of the family unit and class-based discussions of chastity. The particular emphasis in the late-nineteenth century on restraint and abstinence was, as Stephen Garton notes, a result of religious ideas combined with 'a flourishing moral reform movement, socialism and Neo-Malthusianism'.[33] The Victorian valorisation of chastity was because of its wider social implications. After puberty females were not thought to be asexual, but only respectable women were thought sufficiently modest to control and hide their sexual feelings. Women who failed to do so lost their claim to respectability and, by extension, to victimhood.

Unchastity was irrelevant in law, but medical jurisprudence texts were vocal about the difference between legal technicalities and practice. Prominent medico-legal writer Alfred Swaine Taylor noted that 'the law protects a prostitute against involuntary connection just as it protects children and chaste women', but Guy and Ferrier stated in *Principles of Forensic Medicine* that 'it is usual ... to endeavour to rebut the charge of rape by alleging previous unchastity – a question on which the medical examiner may have to express an opinion'.[34] Unchaste victims were actually rare at trial because many such cases were rejected in early stages of the judicial process; they generally only reached trial if the case involved compelling witness testimony. Not a single case reached trial involving a prostitute as complainant, and the grand jury declared 'no bills' for five (of ten) cases in which girls were found to have been previously unchaste. In only five cases did medical witnesses reach trial who explicitly interpreted the absence of a hymen, without *carunculae mytriformes*, as a sign of a girl being accustomed to sexual intercourse. As only one complainant had alleged incest involving repeated long-term sexual contact, medical evidence about the absence of virginity was generally a negative comment on a complainant's character rather than the corroboration of a charge. In all ten cases involving unchastity, or at least medical evidence thereof, the girls were aged between ten and 17. This age grouping was significant. At puberty, female unchastity was increasingly normalised but still frowned upon; as Elizabeth Blackwell wrote in 1894, '[i]t is impossible to reprobate too strongly the false views of physiology held by those who make no distinction between the natural healthy growth of these functions and their abuse'.[35] The group also notably excluded both the very young and adult women. In excluding older, including most married, women from such discussions, medical witnesses revealed the social dimensions of their testimony. Married women were the most likely to have lost the signs of virginity and yet were absent from medical testimony on virginity both at magistrates' courts and at trial. Unchastity was only a medico-legal concern when it was a moral concern: at (or close to) puberty.

The limited number of cases of unchastity that reached trial was, in part, because magistrates dismissed many of them; fortunately, some of these cases can be pieced together from newspaper records. In October 1862, *Lloyd's Weekly Newspaper* reported on evidence given at a Middlesex police court for an alleged rape on 'Esther Whiting, a precocious-looking girl, between 14 and 15 years of age'.[36] The newspaper noted that the medical witness testified:

> [T]hat the girl had not been violated recently, but at a more distant period ... The girl, it was said, had made no great resistance, had followed him into the kitchen, made no outcry nor attempt to escape, and accepted half-a-crown. All the facts were against her having spoken the truth, and conduced to the belief that there was not the slightest evidence against the defendant, whom [the magistrate] discharged.[37]

The girl's lack of resistance was legally relevant, as she was above the age of sexual consent. According to the newspaper, medical evidence about her earlier loss of virginity also played a part in the decision to dismiss this case. Whiting's 'precocious-looking' appearance was inextricably woven with precocious sexual behaviour, neither of which fitted stereotypes of innocent victimhood. This precocious and fallen girl represented dangerous working-class sexualities rather than a stereotype of childish or feminine victimhood, therefore her case never reached trial.

When another magistrate dismissed a case on the basis of the complainant's character, this time in Clerkenwell in 1868, *Reynolds's Newspaper* reported that:

> [T]he surgeon who had examined the prosecutrix was then recalled, and gave such evidence as left no doubt that the prosecutrix could not have been so innocent as she had represented herself to be. (Magistrate) Mr. Cooke said no jury would convict on such evidence, and he should discharge the prisoner.[38]

This magistrate recalled the medical witness for the specific purpose of finding out more about the girl's 'innocence', which referred implicitly to her sexual history. Despite the legal irrelevance of such questions, the courts actively encouraged the blurring of medical, moral and legal boundaries. The medical evidence in this case fed into and corroborated a much broader set of concerns about social class, respectability and honesty. The magistrate apparently did not doubt that the lodger had had sexual intercourse with the girl but, on the basis of her past character, believed her to have consented. The complainant's age was not specified, but she was likely to have been a young adult like many servant girls of the time. As a young female servant, the complainant's honesty was inherently in doubt compared with a household lodger. That she was unchaste and had stolen from a past master only added to questions about her character.

These two magistrates made very similar statements that 'it was quite clear no jury would convict' and 'no jury would convict'. Such comments anticipated a jury's reaction to evidence and, in so doing, drew as much on social stereotypes as on legal requirements. When magistrates presented juries with cases that matched contemporary models of 'real rape', in which strangers attacked chaste victims with force, they implicitly reinforced such beliefs. Vanessa Munro and Liz Kelly note the continued existence today of such 'vicious cycles' by which courts anticipate and perpetuate stereotypes of victimhood.[39] 'A prosecutor,' they note, 'in determining whether a case should be forwarded to trial, pre-judges the jury's beliefs about rape and is more likely to forward cases that are consistent with rape myths, leading to the reinforcement of said myths'.[40] Newspapers should also be understood as part of these so-called 'cycles' as, by selectively reporting cases, they shaped jurors' expectations about the evidence that they would and should hear in court; implicitly, newspapers that identified trends in the types of cases that were successful (or otherwise) were self-perpetuating.

Grand juries also often dismissed cases before trial as part of this 'vicious' cycle. They occasionally even did so for cases in which boys were found to have experience of same-sex acts. In a case of alleged indecent assault on a 12-year-old boy from 1897, for example, a general practitioner called in by Devon police testified that the complainant 'had been practicing sodomy for some time, there were two small fissures inside the anus and the anus was easily dilatable'.[41] Despite the legal irrelevance of chastity in general and consent for a boy of this age, the grand jury rejected this case before it reached court. As a boy of 12 was neither a child nor a man, the complainant in this case filled the same dangerous space in the courtroom as the precocious girl. Males who engaged in sodomy also had been connected with false claims and blackmail, both as perpetrators and victims, since the eighteenth century.[42] The widespread belief that habitual sodomites were untrustworthy is indicated by their complete absence as complainants at the south-west Quarter Sessions and Middlesex sessions. Cases involving boys accustomed to sodomy and unchaste girls were commonly dismissed before reaching trial.

From those few cases that reached court, it seems that medical witnesses discussed unchastity with increasing frequency when complainants reached the normal age range for puberty. In one case from 1854 that was passed forward for trial on the basis of other testimony, a surgeon examined a 16-year-old complainant and found that '[i]ntercourse has been had with her, if it had been this morning it was not the first time'.[43] A number of local London papers, including the *Daily News* and *Morning Post*, printed the same account of this Middlesex trial. They noted that the surgeon's evidence 'tended to throw considerable discredit on the girl's account of the assault' and reported that the prisoner was acquitted.[44] Although the newspapers did not address the subject of her prior unchastity directly, their account of the medical testimony implied a link between her sexual history and the general trustworthiness of her account. Medical practitioners were particularly concerned about working-class girls for whom both physical and behavioural precocity could apparently be the norm. In 1878, when 13-year-old Emily Bell complained of an assault by a fellow servant, the medical witness testified that:

> I examined her and in this case there were only the vestiges of the hymen. I am of opinion she is not a virgin. There were no marks of violence or bruises of any kind and if penetration had taken place the day before for the first time a very serious swelling and injury would

have been visible. There was nothing to shew [sic] a recent rupture of the hymen. Connection might have taken place the day before. There are no appearances on the chemise produced of blood as there would be in a rupture of the hymen.[45]

This medical witness did not deny that an assault had occurred the day before. Officially, he cited her prior unchastity as an explanation for the lack of swelling and injury. The moral dimensions of this testimony were, however, self-evident. It connected to the defence case that, according to *Lloyd's Weekly Newspaper*, 'the girl was under notice to leave for her forward conduct and story-telling, and … prosecutrix had previously asked the prisoner to go to bed with her'.[46]

Servant girls like Emily Bell were a focus of Victorian anxieties about unchastity, particularly among medical authors who blamed servants for introducing young girls and boys to masturbation. Servants embodied concerns about uncontrolled youth, the close living proximity of different sexes and working-class sexualities at puberty.[47] Dr William Acton's mid-century study identified servant girls as the social group most likely to have illegitimate children and, despite the cultural trope of masters seducing their young servant girls, most of this illegitimacy was the result of consensual relations between servants and others of the same social class.[48] Linking to such concerns, the body of the servant girl often raised questions about chastity and consent in court. In 1885, in an unusual case that involved a servant girl making complaint against both her master and mistress, the medical witness testified that:

> The vagina was contused and slightly lacerated and there were bruises on the left thigh and just below the knee and the left breast was bruised … I noticed on that examination that the hymen was ruptured. I should think it had been ruptured within 10 days or a fortnight. The hymen is sometimes ruptured by other means than by the inserting of a foreign body. It may be ruptured by a woman being thrown down. I don't think the violence as detailed in evidence in this case would hardly cause the rupture of the hymen. In my opinion the girl's hymen had been ruptured before Sunday the 11th … (cross-examined) Hearing the girl say that she had been sleeping with a young man for a week before the 11th I am not surprised to find the hymen ruptured.

Again, this medical witness did not use the girl's prior unchastity to deny the offence. He acknowledged the possibility that the hymen had been previously ruptured without sexual intercourse, although under cross-examination raised questions about her contact with a young man. Like many other servants, she was also from outside the UK. As a Hungarian she did not come from a country associated with heat and sexuality but, speaking no English without an interpreter, foreignness left her body vulnerable during examination; she could not easily speak to or explain its signs. In this case, other witnesses testified to her character and the man in question emphasised that they had had no sexual contact. However, medical testimony about the hymen's absence had opened space for these questions in the first instance.

Sexual forensics also positively reinforced models of chaste victimhood by highlighting the previous presence of virginity. In one such Middlesex case involving suspected carnal knowledge of an 11-year-old girl in 1869, which was a misdemeanour at this time, a physician testified to having found a 'recent partial rupture' of the girl's hymen.[49] When questioned by a magistrate, he stated that 'I believe that up to within 48 hours of my examination of this child, she was a virgin'.[50] The highest member of the court questioned this physician on the subject of the girl's chastity, indicating its perceived importance at trial. Similarly, when 13-year-old Annie O'Brien was apparently assaulted in London's Cremorne Gardens in 1875, the medical witness emphasised under cross-examination that 'I should say that she'd certainly not been in the habit of cohabiting with men'.[51] Even in denying unchastity, however, medical testimony demonstrated a growing *need* to engage with questions of character as complainants neared the age of puberty. A new interest in the prior character of girls was evident in cases involving girls from around the age of ten upwards.

The value placed on female chastity was related to a trend, seen in criminal trials for a range of offences, for a complainant's perceived 'respectability' to be situated against that of the accused. However, comments about 'respectability' must be considered more closely. As Alison Phipps notes:

Arguments in this area tend to be ahistorical and atheoretical, focusing on how individual sexual reputations are measured against stereotypical constructions of feminine behaviour, with little attention paid to where such stereotypes come from and in whose interests they operate. A key problem is that respectability is seen as a paradigmatically feminine characteristic rather than as a concept marked by both gender and social class.[52]

This call to arms is compelling, but not straightforward. It is difficult to locate the origin of models of chaste female victimhood because these stereotypes had such long roots, ranging from Christian morals to humoral medicine. Female sexuality came under scrutiny in the nineteenth century as the result of a fierce adversarial system in court and a general growth of moralistic Protestantism that were not necessarily echoed elsewhere. Carolyn Conley notes that 'unlike in England where decisions in cases of sexual assault depended on the perceived moral character of the parties involved, Irish courts convicted in rape cases regardless of the status and relationship of the victim and the accused'.[53] Despite the strict moral codes of Roman Catholicism, Conley finds that magistrates considered it against 'Christian feeling' to allege that women deserved an assault on the basis of their behaviour.[54] In contrast to the relative gender equality of the Irish courts, the strict religious ideals of late-nineteenth century England underpinned a middle-class culture that was intolerant of pre-marital sex.[55]

As Phipps notes, many contemporary anxieties about respectability and unchastity were also grounded in contemporary class hierarchies. Contrary to widespread myths about the real and imagined asexuality of Victorian women, contemporaries were actually highly anxious about disorderly working-class sexualities.[56] In the late-nineteenth century, respectability was only a 'paradigmatically feminine characteristic' for middle-class females. A heightened concern about blackmail, after the 1885 Criminal Law Amendment Act raised the age of sexual consent, also fuelled class-based concerns about female respectability. Medical practitioners and jurors alike were on heightened alert for false claims by the turn of the century, as a result of which unchastity became a marker of general character including the propensity to blackmail and lie. Phipps's comment on the importance of recognising respectability as a 'concept shaped by gender and social class' also requires the addition of age. With the growing gap between biological and social puberty over the course of the nineteenth century, contemporaries paid increasing attention to puberty as a period of uncontrolled sexuality. The issues of puberty, chastity, sexual maturity and the perceived propensity of women to lie were inextricably bound together within and beyond medical thought.

Testimony on unchastity, particularly of pubescent or adult females, had an important role in shaping verdicts. Juries, as magistrates anticipated (albeit in part as a self-fulfilling prophecy), showed little sympathy for complainants in such cases. Not only were many cases dismissed before trial, but newspapers often drew links between sexual character, history and trial outcomes. Medical witnesses had unique access to the body, but such testimony did not need to come from medical witnesses to be effective; an illegitimate child, for example, provided public evidence of an apparent lack of respectability. In 1902 one Devon jury acquitted a prisoner without hearing the case for the defence in an indecent assault case, because the 'young woman ... admitted in cross-examination that she was the mother of two illegitimate children'.[57] In another Devon case from five years earlier, involving a 20-year-old complainant, a local newspaper had similarly reported that the jury convicted the prisoner only on the minor charge of common assault because 'it transpired that the prosecutrix when 16 years of age was delivered of a child ... the girl was not a modest girl'.[58] The notion that a 16-year-old girl should show 'modesty' had clear points of comparison with medical literature and advice texts about the expected development of a 'modest' or a 'retiring' nature at puberty, which would cover up any nascent sexuality. Such similarities indicate the presence of common frameworks of thought rather than the impact of medicine on the judiciary or *vice versa*.

Juries seized upon evidence of female unchastity at or after puberty, but often rejected sexual play or self-abuse as explanations for the genital injuries of younger girls. In the Middlesex case cited above, in which the medical witness testified that a 'child's finger' might have lacerated the hymen of a nine-year-old girl, the jury gave this possibility barely a second thought. According to a London newspaper the jury 'immediately' convicted the prisoner, despite there being no witnesses to the alleged assault.[59] In cases involving children before puberty, the courts focused instead on sufficiently punishing prisoners for causing the *loss* of chastity. When a prisoner was found guilty of an indecent assault on a nine-year-old girl in 1874, for example, *The Times* reported that 'the judge sentenced the prisoner to be imprisoned and kept to hard labour for nine calendar months, remarking that it was a most abominable case, and that the child was probably ruined for life'.[60] Such comments referred to the ruin – physical, mental and moral – of young girls, who could not recover their chastity once 'fallen'. Newspapers reported that when two charges of indecent assault, on girls aged nine and ten, came before one Middlesex judge in the 1870s he 'remarked on the terrible effects which might flow from corrupting the mind of so young a child, and sentenced him to six months imprisonment with hard labour' in one case and, in the other, 'said that the contamination of the mind of a young girl was a serious offence, and sentenced the prisoner to be kept at hard labour for three months'.[61] Although the latter case was apparently mitigated by the prisoner's previous good character, judges considered the sexual purity of young girls to be an aggravating factor during sentencing. That the same judge made these comments twice indicates that there may have been some personal element to such concerns, although they seem to have been broadly representative of judicial practice and drew upon wider concerns about 'pollution' and the 'fallen' female.

The question of chastity remained central to criminal trials throughout the late-nineteenth and early-twentieth centuries. The middle classes increasingly emphasised that men, as well as women, should demonstrate sexual restraint, but this attitudinal shift did not mean that the 'fallen woman' lost her taint. Moral concerns about female sexual behaviour were consolidated rather than challenged over the period, in part due to changing law on female age of consent that apparently left men vulnerable to blackmail. C. Graham Grant's textbook for police surgeons, published in 1907 and 1911, emphasised that '[w]hether the victim is chaste or otherwise has no bearing on the legal aspect of the case, although it may influence the minds of the jury'.[62] Grant's distinction between the 'legal aspects' of a case and the jury's response is significant. It indicates that issues such as character could be irrelevant in law but both permitted and influential within the courtroom. Chastity and character, particularly of girls at puberty and in adulthood, clearly fed into the responses of judges and juries at trial; a stereotypical 'victim' was chaste, therefore magistrates passed cases forward for trial that reflected and reinforced such stereotypes. When the very young were morally 'polluted', in contrast, judges and juries passed down heavy punishments in response to the perceived innate purity of these victims before an assault.

A complainant's sexual history influenced trial verdicts, irrespective of whether evidence came from a medical or lay witness, because of its wider social and moral resonance. However, medical witnesses were able to provide more evidence on the topic because of their unique access to the female body. They were also the only witnesses who could speak legitimately about chastity in court, because a woman's chastity was legally irrelevant but necessary to explain bodily signs. Medical evidence about genital signs of unchastity, or at least the little of this testimony that was not dismissed before trial, spoke to wider social anxieties rather than being influential as 'expert' testimony. Shared frameworks of middle-class thought shaped both the nature of medical testimony and how it was received and interpreted at trial. The courts tolerated such moralistic testimony because it connected to other tropes of courtroom scripts, particularly the growing social and judicial trend for victim blaming. Garthine Walker attributes the growth of victim blaming in the nineteenth century to the new adversarial system, but it also came as part of the Victorian middle-class emphasis on the merits of self-control for both genders.[63] Contrary to popular current-day stereotypes of Victorian female passivity, medical and social texts made constant reference to the dangers of uncontrolled female sexuality at puberty in the absence of adequate 'modesty'. Female and male sexualities were both subjects of concern, apparently needing control and regulation. Medical testimony on unchastity was part of a wider genre of evidence that related to concerns about working-class morality, including living arrangements, alcohol consumption and female conduct with male companions. The prevalence of such moralism, both within and beyond medical testimony, aids an understanding of the positive reception of medical testimony on chastity and sexual character despite its legal irrelevance.

It was only in the twentieth century that expressions of sexuality came to be seen as a normal part of femininity with the decline of the legacies of Evangelicalism, the rise of sexology and a so-called 'sexual revolution'. As pre-marital sex and illegitimacy became increasingly normalised in the late-twentieth century, the expectation declined that an unmarried female would be a virgin. The growth of bacteriological methods and 'rape kits', including DNA analysis in some cases, means that the need for medical witnesses to speak about a complainant's sexual history in order to interpret genital signs has also declined. However, echoes of Victorian and Edwardian attitudes to unchastity are still evident in cases in which a complainant's record of promiscuity – particularly including previous sexual engagement with the accused – finds its way to court. As one *Guardian* journalist noted in 2007: '[d]espite legislation introduced in 1999 to restrict defence barristers from raising a complainant's sexual history in court, judges all too often allow them to get away with it. I have witnessed defence lawyers badgering women with questions about their sexual activity while judges and prosecutors do nothing to stop them'.[64] In the 2000s a number of police forces made it a priority to increase conviction rates for sexual crimes against prostitutes, and had some success in doing so. However, that such measures were necessary indicates that sexual character continues to shape responses to sexual crime.[65]

Case Analysis: Unchastity, 1885

The Case: This pre-trial statement contains evidence relating to the character and chastity of a physically mature young woman. It does not address issues particular to young complainants, such as masturbation or precocity, but has great significance for some other important questions such as the implications of unchastity for the question of consent. The case highlights the role of medicine in propagating concerns about the character of post-pubescent females. It is a valuable case for demonstrating the lack of sympathy for women who did not fulfil middle-class ideals of femininity.

The Prisoner: Richard O'Brien and Robert Phelps, no details provided.
The Complainant: Alma Seldon, aged 18.
The Complaint: Attempted rape (consent a defence).
The Pre-Trial Statement: Depositions taken April 1885, in Somerset.[66]

Alma Seldon: I am 18 years of age and am a singlewoman, living in Union Street, Bridgwater ... I know the two prisoners. I did not know them before yesterday when I spoke to them in the afternoon just above the Bridge in Fore Street, Bridgwater. Sidney Cox was with them at the time. I knew Sidney Cox before and have kept company with him when I lived in Bridgwater about 15 months ago, I have since then been away. In the presence of the two prisoners I made an appointment with Cox to meet him at 6 o'clock the same evening on the Bridge in Fore Street. I went to the place at the time appointed and O'Brien spoke to me before Cox who didn't like it. I shook hands with O'Brien. Cox said 'you're a nice young woman to go for a walk with another when you promised me'. O'Brien asked where I was going and I said 'for a walk'. When I spoke to Cox after that and asked him if he was going with me he said 'no, after you have made an appointment to go with another young man instead of going with me' ... I and O'Brien went on to the Bath Road, Cox would not come ... O'Brien attempted to pull my clothes up. I said 'if you don't be quiet I'll give you in charge of the police'. I walked away back towards Bridgwater to another gate, he then took me hold by the Ulster and threw me down inside the gate, he overtook me when I was walking on. He committed the assault first inside the gate ... and as he pushed me the gate flew open and I fell down, the prisoner got down with me and lifted up my clothes and had connexion with me. When he threw me down I called for help and he called the other young man to hold my arms ... Phelps came in the field and held me down. I couldn't get away from O'Brien because the other was holding me down while he was having connexion with me. I did not consent to O'Brien having connexion with me, it was against my will. (further saith) ... I did not consent to what took place. I denied him three times. The Policeman (Mr Gould) asked him what he was doing. When the Policeman came up I was on my back and O'Brien was on me and Phelps was at the side of me holding my leg and arm and when he saw the policeman he ran across the fields. The policeman took O'Brien by the collar and he told the police he had been keeping company with me and I said he had not ... (cross-examined by Mr Cook, solicitor for both prisoners) ... I was in service with Mrs Chedzoy in Church Street, Bridgwater. I left at a days notice. I got no notice. We fell out ... I was engaged to Cox once, it was broken off 15 months ago before I left Mrs Chedzoy's, he broke it off. I had nothing to do with Cox for about 15 months until Sunday last. I met Cox by accident in Fore St on Sunday afternoon about ½ past 4 and agreed to meet him on Bridgwater Bridge about 6 the same evening. I went to the Bridge about 6 o'clock Sunday evening, then it was that I saw the two prisoners on the bridge. I shook hands with O'Brien and Phelps on the Sunday afternoon, I had never seen them before. [...]

Sidney Cox: I live in the Bristol Road, Bridgwater and am a shipwright. I saw the prosecutrix in Fore Street on Sunday afternoon last about 20 minutes. I had some conversation with her. O'Brien was present, no one else. She asked me if I was coming out at night. I said 'yes, I'll be on the bridge about 6 o'clock'. She said 'alright then'. I wished her good afternoon. I was at the Bridge about a quarter to 6 o'clock. I met with O'Brien and Phelps there. About 6 o'clock Alma came up. She didn't see me. I had my back towards her. I was looking in over the rails and O'Brien spoke to her and went and shook hands with her and he asked her where she was going and she said 'for a walk'. O'Brien asked her where they should go and with the same I turned round and she saw me. She made no reply to O'Brien. I told her since she had made an appointment to go along with O'Brien she had better go with him and I would stop back. She said 'I don't want to go with O'Brien, I would sooner go with you'. I told her I shouldn't think of going with her after promising to go with O'Brien ... O'Brien said 'shall I go?' and she said 'come on then'. I and Phelps went up to Fore Street and we turned round and walked down the town over the bridge and down Eastover ... we passed by O'Brien and prosecutrix, who were stood against the gate with the umbrella up over them. We did not speak to them but went on a quarter of a mile and turned round and came back again. When we came back they were at the next gate towards Bridgwater, about 5 minutes walk from the gate where I first saw them. They were stood in the same position as they were at the gate where we had first seen them and the umbrella was up.

Neither of us spoke to them nor did they speak to us. We walked on about 200 yards, Phelps stopped before I did, not far from the gate. I looked round and could not see Phelps – then I heard one scream, nothing but that one. I then walked straight on about another 100 years. The scream appeared to come from back along, not from the Bridgwater direction, I could not tell whose voice it was. I stopped, then I turned round and walked back towards the place where I heard the scream, then I passed O'Brien and the policeman ... (cross-examined by Mr Cook, solicitor for both prisoners) We were at the bridge from 6 o'clock to half past seven. There was laughing and from going on then between all parties. When prosecutrix had my cane she might have poked me with the cane about half a dozen times about the breast. I did not see her poke the other two. She was poking me with the cane when John O'Brien came across the first time ... I had passed the second gate about 50 yard when I passed the policeman who was going towards the gate, up to that time I had not heard any scream. The moon was just getting up. If any person had been walking in the road or pathway they must have seen them, the cottages were about 50 yards from gate no. 2. I believe they are occupied. When I and Phelps passed them at gate no. 2 they were standing against the gate face to face sideways against the gate.

William Gould: I am a Police Constable stationed at Bawdrip. On Sunday night as I was going from Bridgwater to Bawdrip I saw a man standing on the footpath at the side of the road a few yards from East Bower Lane. Directly as I passed this man I heard a female's voice cry out 'oh do come and help me'. The sounds appeared to come from over the hedge about 50 yards beyond where I was towards Bawdrip, I ran to the place, when I got to a gate I saw someone lying on the ground inside the gate. I heard the person cry out two or three times 'oh do come and help me.' I saw a man in a stooping position keeping the prosecutrix down on the ground, he looked up and saw me, then he ran away across the fields. I can't say who that man was. I went inside the gate to the spot where they were lying and I found the prisoner O'Brien on the prosecutrix. As I was stepping in from the gate towards them, I said 'holloa what are you about here.' I took O'Brien by the collar. I said 'this is a pretty sort of a game isn't it what have you been doing' he said 'oh not much it's the girl I've kept company with'. When he got up I asked him his name and I believe he gave his name as Cox. I know he gave the address Barclay Street. I then asked the prosecutrix her name, she told me her name was Seldon. Before I asked prosecutrix her name she said 'I'll give him in charge' nothing more ... I then went out in the town and met the other prisoner and asked him if his name was Phelps and he said yes, I said 'haven't you been on the Bath road tonight?', he said 'yes', I then said 'and in company with O'Brien?', he said 'yes'. I asked him if he was not in company with O'Brien and a girl inside a gate on the Bath Road, he answered 'yes it's no use telling any lies about it I shall tell the truth', he said 'I saw them in over the gate I went in and said get off Dick and let her go'. He then said 'I put my hands and caught her by the arms to heave Dick off'. I said what made you run away when you saw me coming to the gate. He said 'I wish I hadn't' ... I have examined the prisoner O'Brien's drawers this morning and found one stain as of blood on the front part of his drawers. There was mud and dirt on O'Brien's trousers on the Sunday night. (cross-examined by Mr Cook, solicitor for both prisoners) ... At the moment I got to the gate the prisoner O'Brien was in motion on the girl and she was struggling with her feet and body. I saw or heard nothing more at the moment I got to the gate. I think O'Brien told me out there that he asked of her and she said yes. The girl said 'I did not tell him yes'. O'Brien then turned round and said 'yes, you did.'

Joseph Vowles: I am Superintendent of Police of the Bridgwater Division. On Sunday night O'Brien was brought to the station by P.C. Gould accompanied by the prosecutrix. In consequence of what I heard I charged O'Brien with committing a rape on the prosecutrix, he said I asked her and she consented. Prosecutrix said 'it's a lie I did not' – O'Brien said you did two or three times at the first gate and they contradicted each other several times and I stopped them.

Henry Marcus Kemmis: I am a Surgeon practising in this town. On Sunday night last the prosecutrix was brought to my house by Supt Vowles about ½ past 10 or a quarter to 11. In consequence of what I was told I examined the prosecutrix. I found her under clothes were stained a good deal but I did not find anything of a recent date. As regards the vagina, there was a slight discharge, very slight, but there were no marks of any violence at all. If a person had been violently ravished two or three hours previous I should have expected to have found some marks of violence. As far as my examination went there was nothing to arouse my suspicions that anything such as rape had taken place. (cross-examined by Mr Cook, solicitor for both prisoners) The ordinary signs of virginity were destroyed and had been destroyed some time before that night, but how long before I couldn't tell, according to appearances some time before. With regard to the stains on the underclothing, they were caused by her monthly courses. It would have been completely impossible for a young man like the prisoner to have connexion with the prosecutrix while she was struggling and kicking without some signs of violence being visible.

Both prisoners said: 'not guilty'

For the Defence

John O'Brien: I am a labourer and live at the old Dock, Bridgwater. The prisoner O'Brien is my nephew. I was on Bridgwater Bridge last Sunday evening between 6 and 7 in company with Robert Slocombe and others. I saw the prosecutrix there also Cox and the two prisoners all four together. The prosecutrix was too familiar, poking Cox with a cane. They were laughing and giggling a long time. I am the father of a family. I told her to go away and let the youngsters alone or she would get into trouble herself or get them into trouble. She made no reply but moved about two yards off. Phelps said ... we can't get rid of her ... I said you nasty dirty little faggot go away and let them alone and I told my nephew Richard to go home, she heard it and said what odds is it to you. I went the other side of the bridge and when I looked round she was fiddling then with a stick. I did not see which way they went, I left them there. [...]

The Verdict: No Bill.

Significance for Sexual Forensics: In this case, it was difficult to deny that some degree of sexual contact had taken place between the girl and at least one prisoner. In addition to the girl's testimony and another witness who heard her scream, a policeman 'found the prisoner O'Brien on the prosecutrix' and apparently saw her struggling against the act. The failure of this case seemingly rested not on a denial of the sexual act, but on the girl's perceived low character, for which evidence came in part from lay witnesses to her over-familiar conduct with the boys and in part from medical testimony about her previous unchastity. The medical evidence in this case was crucial for tipping the scales against the complainant, as lay witnesses had already provided evidence that both corroborated and disputed the charge.

The medical practitioner provided solid evidence of the girl's lack of respectability in the form of evidence about her unchastity. The prisoners' solicitor encouraged this medical testimony about the girl's sexual history as, the complainant being 18 years of age and previously engaged to be married, there was no reason to assume that she was physically pure. From the perspective of the prisoners' solicitor, the medical witness provided the perfect answer: evidence that Alma Seldon had engaged in sex 'some time before' reaching the age of 18. Seldon's behavioural maturity was matched by her physical maturity, as the medical witness noted stains from her 'monthly course' (menstruation). Overall, there was little in the medical testimony to indicate that the complainant fulfilled the qualities generally ascribed to female victimhood: chastity alongside physical / mental immaturity or frailty. The medical witness did not dispute that sexual contact might have taken place between the complainant and prisoner, but emphasised that there were no marks of having been '*violently* ravished' (added emphasis). The absence of signs of virginity meant that there would be no evidence that penetrative sex had occurred unless accompanied by violence. In the light of the complainant's sexual history and other witness testimony, this medical testimony indicated that she might have engaged in a consensual sexual act, which she then denied upon being caught, but that she had not been the victim of a non-consensual encounter.

In testifying that it would be 'impossible' to rape a struggling woman without evidence of violence, the medical witness did not necessarily dispute the policeman's observations about seeing the complainant struggle. He implied either that the prisoner did not complete the act or, drawing upon wider contemporary 'rape myths', that the complainant might have feigned resistance rather than giving 'real' resistance; as the subsequent chapter on resistance shows, the rape myth that ' "no" might really mean "yes" ' was widespread at the time of this alleged offence. Medical testimony tapped implicitly into wider social thought that an unchaste female was more likely than a chaste woman to consent to sexual contact with any man. If the medical evidence about the 'impossible' nature of rape with a struggle could have been interpreted in two ways (as an attempted but unsuccessful assault, or as the girl's feigned resistance), it seems significant that the grand jury opted for a 'no bill' instead of a lesser charge. This outcome indicates that the solicitor's attempt to draw out implicit links from medical testimony, between unchastity and consent, were successful. Irrespective of the fact that Seldon's previous unchastity may have involved only the man to whom she was engaged, a question not addressed in court, the medical evidence confirmed lay testimony about her over-familiar nature. To quote the prisoner's uncle, it was widely believed that such girls got both themselves and their easily-led male peers 'into trouble'.

Notes

1. Johann Ludwig Casper, *Handbook of Forensic Medicine*, trans. from 3rd edn by G. W. Balfour, vol. 2 (London: New Sydenham Society, 1862), pp. 279–80. Emphasis in original. For contemporary examples of cases involving pregnant women with intact hymens and virgins with congenitally absent hymens, one of which was provided by an Exeter-based correspondent, see: 'Notes, Short Comments, and Answers to Correspondents', *The Lancet*, 3 November 1888, 898–900, p. 899; 'Physical Signs of Virginity', *The Lancet*, 15 July 1865, 84, p. 84; and Alfred Swaine Taylor, *A Manual of Medical Jurisprudence*, 8th edn (London: J. & A. Churchill, 1866 [1844]), pp. 602–05. For an extensive discussion of the value and problems of the

hymen as a sign see 'The Physical Signs of Virginity', *British Medical Journal* (*BMJ*), 5 January 1895, 27, p. 27. For a historiographical discussion of the problematic nature of the hymen in nineteenth-century medico-legal thought see Ivan Crozier and Gethin Rees, 'Making a Space for Medical Expertise: Medical Knowledge of Sexual Assault and the Construction of Boundaries between Forensic Medicine and the Law in late Nineteenth-Century England', *Law, Culture and the Humanities* 8 (2012), 285–304, pp. 301–31.
2. Sarah Toulalan, ' "Unripe" Bodies: Children and Sex in Early Modern England' in *Bodies, Sex and Desire from the Renaissance to the Present*, ed. Kate Fisher and Sarah Toulalan (Basingstoke: Palgrave MacMillan, 2011), 131–50, pp. 132, 141; Louise A. Jackson, 'Child Sexual Abuse and the Law: London 1870–1914', unpublished doctoral thesis, Roehampton Institute (1997), p. 180; Louise A. Jackson, *Child Sexual Abuse in Victorian England* (London: Routledge, 2000), p. 6.
3. London, London Metropolitan Archives (LMA), Pre-Trial Statements, Robert Thornton tried at the Middlesex Sessions on 21 December 1876 for indecent assault, MJ/SP/E/1876/027.
4. London, LMA, Pre-Trial Statements, Alfred Boydry tried at the Middlesex Sessions on 20 August 1877 for indecent assault, MJ/SP/E/1877/019.
5. Gloucester, Gloucestershire Archives (GA), Pre-Trial Statements, George Norman Robbins tried at the Gloucestershire Quarter Sessions on 18 March 1906 for indecent assault, Q/SD/2/1906.
6. 'Gloucestershire Quarter Sessions', *Bristol Daily Mercury*, 28 June 1906, 7, p. 7; Gloucester, GA, Pre-Trial Statements, Robbins, Q/SD/2/1906.
7. For boys, the link between precocity, delinquency and crime related to theft rather than prostitution. For gendered links between precocity and crime see Shani D'Cruze and Louise A. Jackson, *Women, Crime and Justice in England since 1660* (Basingstoke; New York: Palgrave Macmillan, 2009), p. 75; Margaret May, 'Innocence and Experience: The Evolution of the Concept of Juvenile Delinquency in the Mid-Nineteenth Century', *Victorian Studies* 17 (1973), 7–29, p. 21; and Larry Wolff, ' "The Boys are Pickpockets, and the Girl is a Prostitute": Gender and Juvenile Criminality in Early Victorian England from *Oliver Twist* to *London Labour*', *New Literary History* 27 (1996), 227–49.
8. William Acton, *Prostitution, Considered in its Moral, Social, & Sanitary Aspects, in London and Other Large Cities with Proposals for the Mitigation and Prevention of its Attendant Evils* (London: John Churchill, 1857), p. 60.
9. Daniel Pick, *Faces of Degeneration: A European Disorder, c. 1848–1918* (Cambridge: Cambridge University Press, 1993), p. 5.
10. Mary Douglas, *Purity and Danger: An Analysis of the Concepts of Pollution and Taboo* (London: Routledge and Kegan Paul, 1978), p. 40.
11. See the Rosina Symes case analysis in 'consent' for an example of a case in which a girl was implicitly considered to be 'precocious' after a second complaint to the courts.
12. Gloucester, GA, Pre-Trial Statements, David Thomas Davis tried at the Gloucestershire Quarter Sessions on 19 October 1887 for indecent assault, Q/SD/2/1887.
13. London, LMA, Pre-Trial Statements, Boydry, MJ/SP/E/1877/019.

14. Taunton, Somerset Heritage Centre (SHC), Pre-Trial Statements, John Henry Hayman tried at the Somerset Quarter Sessions on 1 July 1903 for indecent assault, Q/SR/812.
15. Even in 'normal' puberty the emergence of the sexual instinct was expected to precede the ability to control it. However, in cases of precocity this disjunction between physical and mental development was deemed even greater and therefore even more concerning.
16. Elizabeth Blackwell, *The Human Element in Sex: A Medical Inquiry into the Relation of Sexual Physiology to Christian Morality* (London: J. & A. Churchill, 1894), p. 34.
17. 'The Week', *BMJ*, 6 February 1864, 161, p. 161.
18. William Acton, *The Functions and Disorders of the Reproductive Organs in Youth, in Adult Age, and in Advanced Life, Considered in their Physiological, Social and Psychological Relations* (London: John Churchill, 1857), p. 55.
19. Steven Marcus, 'Introduction' [1975] in Sigmund Freud, *Three Essays on the Theory of Sexuality*, trans. and ed. James Strachey (New York: Basic Books, 2000 [1905]), xxxi–liii, p. xxxi.
20. 'Reviews: Psycho Analysis', *BMJ*, 5 July 1913, 23–24, p. 23. For some more positive discussions of Freud's ideas see J. A. Ormerod, 'The Lumleian Lectures on Some Modern Theories Concerning Hysteria', *The Lancet*, 9 May 1914, 1299–303, p. 1302; 'Puberty and the Neuro-Psychic System', *BMJ*, 3 June 1913, 1337, p. 1337.
21. London, LMA, Pre-Trial Statements, William Boddington tried at the Middlesex Sessions on 24 August 1865 for indecent assault, MJ/SP/E/1865/016.
22. Ogston, *Lectures on Medical Jurisprudence*, p. 104.
23. See, for example, T. S. Clouston, 'Puberty and Adolescence Medico-Psychologically Considered', *Edinburgh Medical Journal* 26 (1880), 5–17, p. 6; 'The White Slave Traffic Bill', *BMJ*, 5 October 1912, 969, p. 969.
24. Self-control was described in terms of the male qualities of 'will' or 'reason' and the feminine traits of 'modesty' or 'reserve'. This gender division is simplistic to an extent, but was prevalent in contemporary medical literature and has been identified in historiography such as Lucy Bland, *Banishing the Beast: English Feminism and Sexual Morality 1885–1914* (London: Penguin, 1995), p. 56.
25. London, LMA, Pre-Trial Statements, George Kimpton tried at the Middlesex Sessions on 28 July 1875 for assault with intent to commit an unnatural crime, MJ/SP/E/1875/013.
26. 'Spermatorrhoea' referred to the involuntary loss of semen and was coined by Claude-François Lallemand in French and translated works of the early-nineteenth century, although drew on a longer history of medical discussions of 'seminal weakness'; see Elizabeth Stephens, 'Coining Spermatorrhoea: Medicine and Male Body Fluids, 1836–1866', *Sexualities* 12 (2009), 467–85.
27. 'The Rapid Heart: A Clinical Study', *The Lancet*, 10 July 1890, 1001–06, p. 1002.
28. See Susan Sniader Lanser, 'Befriending the Body: Female Intimacies as Class Acts', *Eighteenth-Century Studies* 32 (1999), 179–98.
29. Robert H. MacDonald, 'The Frightful Consequences of Onanism: Notes on the History of a Delusion', *Journal of the History of Ideas* 28 (1967), 423–31, p. 431; Mark Breitenberg, *Anxious Masculinity in Early Modern England* (Cambridge: Cambridge University Press, 1996), p. 53; Michael C. Schoenfeldt, *Bodies and Selves in Early Modern England: Physiology and*

Inwardness in Spenser, Shakespeare, Herbert, and Milton (Cambridge: Cambridge University Press, 1999), p. 83.
30. Sally Shuttleworth, *The Mind of the Child: Child Development in Literature, Science and Medicine, 1840–1900* (Oxford: Oxford University Press, 2010), p. 61.
31. J. H. Badley, 'How the Difficulties in Teaching Eugenics May Be Overcome', *Eugenics Review* 5 (1913), 12–18, p. 12.
32. Michael Anderson, 'The Social Implications of Demographic Change' in *The Cambridge Social History of Britain 1750–1950*, ed. F. M. L. Thompson, vol. 2 (Cambridge: Cambridge University Press, 1990), 1–70, p. 28.
33. Stephen Garton, *Histories of Sexuality* (London: Equinox, 2004), p. 106.
34. A. S. J. Taylor, *The Principles and Practice of Medical Jurisprudence*, ed. Thomas Stevenson, 6th edn, vol. 2 (London: J. & A. Churchill, 1910 [1865]), p. 116; William A. Guy and David Ferrier, *Principles of Forensic Medicine*, 5th edn (London: H. Renshaw, 1881 [1844]), p. 61.
35. Elizabeth Blackwell, *The Human Element in Sex: A Medical Inquiry into the Relation of Sexual Physiology to Christian Morality* (London: J. & A. Churchill, 1894), pp. 26–27.
36. 'Unfounded Charge', *Lloyd's Weekly Newspaper*, 26 October 1862, 7, p. 7.
37. 'Unfounded Charge', p. 7.
38. 'Charge of Indecent Assault', *Reynolds's Newspaper*, 2 February 1868, 5, p. 5.
39. Vanessa E. Munro and Liz Kelly, 'A Vicious Cycle? Attrition and Conviction Patterns in Contemporary Rape Cases in England and Wales' in *Rape: Challenging Contemporary Thinking*, ed. Miranda Horvath and Jennifer Brown (Cullompton: Willan Publishing, 2009), 281–300.
40. Munro and Kelly, 'A Vicious Cycle?', p. 295.
41. Exeter, Devon Record Office (DRO), Pre-Trial Statements, Joseph Groves not tried (no bill) at the Devon Quarter Sessions in April 1897 for indecent assault on a male, QS/B/1897/Easter.
42. Angus McLaren notes that '[a]s early as the 1730s references were made to the organized blackmail of sodomites in London ... Between 1885 and 1900 there was a surge in the reportage of attempts at blackmail'; Angus McLaren, *Sexual Blackmail* (Cambridge, MA: Harvard University Press, 2002), pp. 13–18.
43. London, LMA, Pre-Trial Statements, Charles Bruno tried at the Middlesex Sessions on 1 May 1854 for indecent assault, MJ/SP/E/1854/015.
44. 'Middlesex Sessions – Yesterday', *Daily News*, 3 May 1854, 6, p. 6. The same account can be found in 'Middlesex Sessions', *Morning Post*, 3 May 1854, 7, p. 7.
45. London, LMA, Pre-Trial Statements, Henry Emeny tried at the Middlesex Sessions on 28 August 1878 for carnal knowledge, MJ/SP/E/1878/027.
46. 'Yesterday's Law & Police', *Lloyd's Weekly Newspaper*, 28 August 1878, 12, p. 12.
47. Fears about 'servant girls' were articulated in medical texts ranging from sexology to more mainstream literature; See Havelock Ellis, *Studies in the Psychology of Sex*, vol. 3 (Philadelphia: F. A. Davis, 1904 [1903]), p. 175; and Athola W. Johnson, 'On an Injurious Habit Occasionally Met With in Infancy and Early Childhood', *The Lancet*, 7 April 1860, 344–45, p. 344.
48. John R. Gillis, 'Servants, Sexual Relations and the Risks of Illegitimacy in London, 1801–1900', *Feminist Studies* 5 (1979), 142–73.
49. London, LMA, Pre-Trial Statements, Joseph Dungay tried at the Middlesex Sessions on 23 December 1869 for carnal knowledge, MJ/SP/E/1869/025.
50. London, LMA, Pre-Trial Statements, Dungay, MJ/SP/E/1869/025.

51. London, LMA, Pre-Trial Statements, Henry Margetson tried at the Middlesex Sessions on 15 November 1875 for indecent assault, MJ/SP/E/1875/022.
52. Alison Phipps, 'Rape and Respectability: Ideas about Sexual Violence and Social Class', *Sociology* 43 (2009), 667–83, p. 675.
53. Carolyn Conley, 'No Pedestals: Women and Violence in Late Nineteenth-Century Ireland', *Journal of Social History* 28 (1995), 801–18.
54. Conley, 'No Pedestals'.
55. Ruth H. Bloch, 'Untangling the Roots of Modern Sex Roles: A Survey of Four Centuries of Change', *Signs* 4 (1978), 237–52; and Edmund S. Morgan, 'The Puritans and Sex', *The New England Quarterly* 15 (1942), 591–607.
56. On the modern myth of Victorian asexuality see Michael Mason, *The Making of Victorian Sexuality* (Oxford; New York: Oxford University Press, 1994).
57. 'Devon Sessions', *Trewman's Exeter Flying Post*, 18 October 1902, 3, p. 3.
58. 'Devon Quarter Sessions', *Trewman's Exeter Flying Post*, 23 October 1897, unpaginated.
59. 'Middlesex Sessions', *Standard*, 25 August 1865, 7, p. 7.
60. 'Middlesex Sessions', *The Times*, 19 June 1874, 11, p. 11.
61. 'Middlesex Sessions', *The Times*, 25 January 1877, 11, p. 11.
62. C. Graham Grant, *Practical Forensic Medicine: A Police-Surgeon's Emergency Guide*, 2nd edn (London: H. K. Lewis, 1911 [1907]), p. 43.
63. Garthine Walker, 'Rape, Acquittal and Culpability in Popular Crime Reports in England, c.1670–c.1750', *Past and Present* 220 (2013), 115–42.
64. Julie Bindel, 'Why is Rape so Easy to Get Away With?' <http://www.theguardian.com/society/2007/feb/01/penal.genderissues> (accessed 1 March 2015).
65. 'Police convict 90 per cent of those who rape sex workers in Liverpool' <http://www.liverpoolecho.co.uk/news/liverpool-news/police-convict-90-those-who-3451454> (accessed 1 March 2015); 'Merseyside police: "Sex workers are vulnerable. We want to protect them" ' <http://www.theguardian.com/society/2010/dec/22/merseyside-police-sex-workers-protect> (accessed 1 March 2015).
66. Taunton, Somerset Heritage Centre (SHC), Pre-Trial Statements, Richard O'Brien and Robert Phelps not tried (no bill) at the Somerset Quarter Sessions in April 1885 for attempted rape, Q/SR/739. Due to the length of the original pre-trial statement, it has been edited down to the most relevant testimony. Some punctuation has been added for clarity, but the meaning remains unchanged.

4
Consent: Violence and the Vibrating Scabbard

'No sane man can believe, that a woman of average height and strength, and not overcome by drugs, could be violated by one man'.[1] These were the words of gynaecologist Lawson Tait in an 1894 article on sexual offences against females. Tait used knowledge of the 'average' woman's body to justify the expectation that she could, and should, fight off an attempted offence. His work suggests that he was more cynical than most of his peers, as he published widely about the prevalence of false claims in rape trials, but was not alone in believing in a normal woman's capacity to resist. *The Medical Press* had published similar comments in July 1890: '[g]enerally speaking, few medical men of experience believe much in rape in the case of a moderately healthy and vigorous woman'.[2] Such claims were grounded in the belief that a man could not achieve penetration of a woman who resisted to her utmost, as articulated through the rape myth that it was 'impossible to sheath a sword into a vibrating scabbard'.[3] As late as 1913 the US Police Surgeon Gurney Williams pointed out in *International Clinics*, published in London and Philadelphia, that mere 'crossing of the knees' would prevent rape.[4]

These claims created a problem for practitioners of sexual forensics. Signs of resistance were necessary for a successful rape conviction, in part because in law the crime was defined as sexual intercourse 'against the will' of a woman. However, according to medical thought, an 'average' woman who truly resisted would have prevented any attempted assault. Medical witnesses overcame this contradictory position by emphasising that, although 'healthy and vigorous' women were capable of resistance, other females might resist to their full capacity and still be assaulted. Girls who had not reached full physical development, females who lacked the psychological capacity to understand a need to resist, and women who struggled to the point of 'insensibility' could all be victims. As with many aspects of sexual forensics, medical evidence about consent and resistance thus reinforced traditional models of 'real rape' that required signs of a physical struggle. It placed a particular onus upon pubescent girls and adult women either to demonstrate this 'real' quality of victimhood or to justify its absence.

The 'double standard' of sexual behaviour was challenged gradually over the course of the nineteenth century, which had a knock-on effect on ideas about consent and resistance. Rape myths relating to sexual consent, such as the belief that in some situations ' "no" might really mean "yes" ', were partly grounded in the assumption that male sexual urges were naturally strong. In this framework, women had a responsibility to demonstrate that they had resisted the advances of men who might not be able to help themselves. As males came increasingly to be held accountable for any failure to control their lust, case law backed away slightly from victim blaming.[5] The law began to acknowledge the psychological dimensions of consent and decreased the onus on female complainants to prove evidence of a lengthy struggle. In practice, however, these shifts had little tangible impact on medical testimony before 1914. Conservative courtroom 'scripts' gave weight to the physical aspects of consent and resistance in the adversarial process and defence strategies. Although judges began to recognise consent and resistance as complex issues, jurors also continued to focus on the physical dimensions of consent and resistance that placed the onus on women not only to resist, but to do so 'to their utmost'. This chapter outlines how, despite increasingly multifaceted definitions of consent in case law and some medical literature, in court medical witnesses and juries continued to support a physical model of consent and resistance; long-held stereotypes were not quickly shaken off by changes in case law.

The belief that women should and could fight off offenders was grounded in a number of age-based assumptions, which meant that questions of consent and resistance were theoretically irrelevant for children. A woman needed to be physically capable of fighting, a basic assumption related to bodily strength. She also needed to know that an act was wrong in order to resist it, a more implicit assumption that related to mental maturity and the capacity to understand sexual acts. The capacity to consent and responsibility to resist were two sides of the same coin, as a woman could only resist if she were able to refuse consent. The law on sexual consent was relatively clear-cut: active dissent was required for a rape conviction to show that a crime was 'against the will', but irrelevant for the felony of carnal knowledge that related to the youngest complainants. The question of consent was relevant only for determining the charge if complainants were in the misdemeanour age group: the misdemeanour of carnal knowledge could be upgraded to the felony of rape with proof of dissent for girls aged 10–12 until 1875, 12–13 until 1885, and 13–16 after 1885.

Between 'consent' and 'resistance' was a third category: submission to a sexual act. This category was the most complex, indicating reluctance or passivity but not necessarily active consent or resistance. Submission had previously had negative connotations, relating to women's failure to resist to their utmost ability, but in the nineteenth century came to be associated with psychological issues such as fear. Medical and legal literature increasingly sought to distinguish submission from consent, albeit with mixed success and – for medicine at least – little impact in practice. Authors of medical jurisprudence highlighted circumstances in which resistance might be normatively absent, even without any threat of physical harm. In the fifth edition of his *Manual of Medical Jurisprudence*, published in 1854, Taylor expanded his discussion of female resistance and the lack thereof. He noted that:

> An eminent judicial authority has suggested to me that, in his opinion, too great distrust is commonly shown in reference to the amount of resistance offered by women of undoubted character. Inability to resist from terror, or from an overpowering feeling of helplessness, as well as horror at her situation, may lead a woman to succumb to the force of a ravisher, without offering that degree of resistance which is generally expected from a female so situated. As a result of long experience, he thinks that injustice is often done to respectable women by the doctrine that resistance was not continued long enough.

Despite Taylor's efforts to break down the binary positioning of 'consent' and 'resistance', this extract spoke only of 'respectable' women. His comments were grounded in a wider social assumption that 'respectable' woman were more likely to feel 'terror … helplessness … horror'. 'Respectable women' implicitly referred to females from the middle and upper classes. However, this category could include respectable working-class complainants who distinguished themselves from society's so-called 'residuum' in the wake of urbanisation and industrialisation.[6] In so doing, they increasingly valorised and emulated middle- and upper-class traits. Respectability came to be interwoven with a particular model of gendered behaviour, that of the higher social classes for whom ideal femininity involved physical and emotional delicacy. It is no coincidence that Taylor included this new point in the 1850s, when increasing emphasis was being placed upon the 'respectable' woman's delicate nerves, sensitivity and passivity. In this gender framework, a woman might be 'helpless' and submit to a sexual act without either active consent or resistance.

Taylor's reference to an 'eminent judicial authority', instead of his own experience in the courts, represented a general medico-legal deference to legal authority on the subject of consent. Taylor emphasised that such issues were outside the medical remit, having written in 1852 that '[a] woman may yield to a ravisher, under threats of death or duress: in this case her consent does not excuse the crime, but this is rather a legal than a medical question'.[7] This deference reflected the fact that most changes in definitions of 'consent', 'resistance' and 'submission' came from the law rather than from medicine. Richard Burn's *Justice of the Peace*, an authority for magistrates, first addressed the issue of submission in 1869 when he noted that '[t]here is a great difference between submission and consent; consent involves submission, but it by no means follows that mere submission involves consent'.[8] Burn cited fear of 'a strong man' or a schoolmaster as reasons why young girls might 'submit' without giving consent. He also noted that an adult woman might submit to the authority of a medical man, without giving active consent.[9] The law increasingly recognised non-consensual 'submission', but only in exceptional circumstances involving power imbalance and/or age difference.

Between 1850 and 1880 a number of case law decisions emphasised that 'submission' could occur without consent, particularly in cases of incest or crimes in which the perpetrator was in a position of authority. In one case law decision from 1872 Justice Brett stated that:

> In cases of criminal assault by a father upon his daughter I have more than once, with the concurrence of Willes, J., told the jury that in the case of a young child and an adult they must consider whether there has been mere submission on the part of the child, known to be merely so by the adult, or whether there has really been consent.[10]

Justice Brett and Justice Willes were both 'commanding' and influential judges of the time.[11] That Willes, a judge known 'for showing mercy to prisoners', encouraged a jury to distinguish between consent and submission in cases of father-daughter incest also indicates that these cases were viewed with particular severity.[12] In an 1870 Central Criminal

Court trial for carnal knowledge of a girl above the age of ten and below the age of 12, Justice Lush similarly directed the jury that:

> On the second count [of indecent assault] you cannot convict if there has been consent, as an assault excludes consent. But consent means consent of will, and if the child submitted under the influence of terror, or because she felt herself in the power of the man, her father, there was no real consent.[13]

These case law decisions meant that, by the 1870s, there was an official distinction between 'real' consent and submission. However, they focused on children rather than on pubescent or adult females. Many of these case law decisions were the result of a particular legal loophole in which consent remained a defence in cases of 'indecent assault' on children but not in trials for 'carnal knowledge'. Judges sought to close this loophole in case law by distinguishing between 'submission' and active consent, the latter of which was apparently impossible for children who lacked understanding of the sexual act. These distinctions became less relevant after the 1880 Criminal Law Amendment Act, in which consent was removed as a defence in cases of alleged indecent assault against girls and boys under the age of 13.

For women over the age of sexual consent, case law of the mid-nineteenth century instead reinforced expectations about active resistance. One 1865 case law decision stated that:

> [O]n a charge of rape, the jury should be satisfied, not merely that the act was in some degree against the will of the woman; but that she was, by physical violence or terror, fairly overcome, and forced against her will, she resisting as much as she could, so as to make the prisoner see and know that she was really resisting to the utmost.[14]

This case law decision demanded evidence that a woman resisted 'to the utmost', rather than only proving that a rape was 'against her will' as required in written law. This approach was grounded in the pervasive contemporary belief that females often feigned a struggle without truly resisting. It placed the responsibility firmly on the female to 'make the prisoner see and know' that 'no' meant 'no', rather than upon the prisoner to gain consent. These ideas were also established in the British Empire through law and medical jurisprudence literature as, for example, Elizabeth Kolsky shows that they were exported to colonial India.[15]

In the late-nineteenth century most case law decisions were unsympathetic to complainants at and above the age of puberty. They reinforced a belief that women should show resistance, even if they were 'weak minded'. Medical testimony sometimes informed these decisions. In a case law decision from 1866, involving an 'idiot' female complainant aged 16, 'the medical man stated she was a fully developed woman, and that strong animal instincts might exist, notwithstanding her imbecile condition'.[16] His ideas corresponded to physiological ideas about the age of bodily maturity and the medical notion that sexual desire without the capacity to control it was a dangerous force. The judge directed that 'a consent produced by mere animal instinct would be sufficient to prevent the act from constituting a rape'.[17] He quashed the guilty verdict on appeal and stated that '[u]pon an indictment for rape there must be some evidence that the act was without the consent of the woman, even where she is an idiot'.[18]

Such ideas were later applied without direct medical involvement, as the courts conflated physical maturity with the ability to consent or resist. In a case from 1873 involving a 14-year-old girl with 'mental incapacity', the case report made no reference to a medical witness, but the judge directed that 'if from animal instinct she yielded to the prisoner without resistance, or if the prisoner from her state and condition had reason to think she was consenting, they ought to acquit him'.[19] It is unclear whether this judge was drawing on the earlier case law decision or on wider contemporary thought about sexual instincts. Either way, the case highlights how medical ideas could be circulated throughout non-medical contexts and in the absence of any direct medical input or control. It was not until the 1885 Criminal Law Amendment Act that all 'imbecile girls' were given legal protection against sexual offences, which marked a legal turn away from physical models of consent and resistance.

In the 1860s and 1870s case law only distinguished between 'submission' and 'consent' in cases involving sexually mature females if they were victims of fraud, despite the law's multifaceted approach to consent in cases involving young girls.[20] It was not until the early-twentieth century that a more nuanced and multifaceted conception of consent became evident in cases involving pubescent or adult women, as part of a general shift towards balancing gender responsibility in cases of sexual crime. In an appeal against an incest conviction involving a girl aged 16 from November 1911, for example, the question of consent was relevant to determining whether the girl was a victim or an accomplice. The prosecution emphasised that '[m]ere submission is not necessarily consent'.[21] When the prisoner's defence argued that 'the girl either consented or she did not. There is no middle course', the judge emphasised that '[t]here is a distinction between submission and permission'.[22] The judge here explicitly rejected the older but enduring notion that, with females who were adult in law, there was 'no middle course' between consent and resistance.

Forensic texts acknowledged these legal debates about the contributing role of factors such as fear and authority to consent or 'submission'. In general, medical writers, however, demonstrated more interest in broad questions about when and why girls gained the *capacity* to consent. An important part of the question of sexual consent from a medical standpoint was the age at which a girl became capable of resisting, in terms of bodily strength. Medical literature presented a particularly flexible view of this question because it looked through the lens of physiology, rather than sexual consent law. Throughout the editions of Taylor's *Manual of Medical Jurisprudence*, he emphasised that only 'infants' were always physically incapable of resistance on the basis of their age; in 1910 he also added 'aged' women to this list.[23]

Many medical writers emphasised that the development of physical strength, although a variable process, came relatively early in puberty and often predated male development; even a relatively young girl might therefore be deemed physically strong enough to resist a male peer. In an article in *The Lancet*, opposing the 1885 Criminal Law Amendment Act, anthropometrist Charles Roberts claimed that 'as far as body growth determines maturity, a girl of 12 years of age (the youngest age referred to in the Criminal Law Amendment Bill) is equal to a boy of 15 years and a half'.[24] Medical dissent from the law on sexual consent was equally explicit in the 1910 edition of Taylor's work, which included a new statement that 'the older the victim, up to our present limit [16], the greater the possibility of such marks on the body and the more the importance to be attached to them when found, or at least to the explanation of how they were produced'.[25] Such distinctions had no clear legal purpose as consent and resistance had long been irrelevant for girls under the age of 13 and relevant only in deciding the charge between 13 and 16. Taylor's comments, grounded in a physiological continuum up to the age of 16, had little relation to this two-tier system.

Medical textbooks acknowledged that consent was multifaceted, not least because understanding a sexual act was a prerequisite for consenting to or dissenting from it. Taylor noted that physical strength would not *always* result in resistance, an observation that he made increasingly explicit over time. Taylor's 1910 *Principles and Practice* noted that 'marks of violence on the *body* of a young child ... are seldom met with, because no resistance is commonly made by mere children'.[26] This lack of resistance was apparently due to a lack of knowledge about sex, with 'normal' and 'respectable' children lacking a capacity to understand the sexual act. Medical books repeatedly emphasised that only precocious girls would have sexual knowledge before puberty.[27] These girls alone would have the understanding and bodily strength both to dissent mentally and to resist a man's sexual advances physically, but such girls were not typical 'victims'. This scientific framework placed an onus on precocious girls to prove the absence of consent, despite its irrelevance in law.

Although medical literature paid attention to different aspects of the capacity to consent, in court the medical witness' remit was to speak just on the 'physical proofs' in front of him. Within sexual forensics, the 'physical proofs' of consent generally meant the absence of a struggle, meaning that in practice resistance came to be equated with the relatively clear-cut sign of bodily violence. Genital injury furnished some evidence of violence, but was thought to be less important as a sign of resistance because it could also apparently occur in a virgin who consented. The focus on a physical struggle, as the key indicator of consent and resistance, differed from medical approaches of earlier centuries. Of particular note is the disappearance from medico-legal texts of the link between pregnancy and consent. Although the idea has been overturned by the eighteenth century, as late as 1815 writers of forensic medicine textbooks emphasised that the 'depressing' nature of a sexual crime would prevent the ability to conceive; by extension, a woman who became pregnant after an alleged rape could not be a true 'victim'.[28] This belief was, however, absent from most mainstream medical jurisprudence by the mid-nineteenth century. In line with new physiological studies of sexual development, medical literature treated pregnancy as a marker of physical maturity rather than consent. However, as with many long-held beliefs, the link between pregnancy and sexual consent lingered in quack literature and even in the works of some professional medical writers. In a lecture published in *The Lancet* in 1883, for example, physician J. Matthews Duncan stated that:

> It is also a popular opinion that desire and pleasure are essential elements in fecundity, and in cases of rape followed by pregnancy, that consequence has been made ground of defence against the charge ... I think it is very nearly certain that desire and pleasure in due or moderate degree are very important aids to, or predisposing causes of, fecundity.[29]

Duncan admitted his lack of evidence for such claims, however, and was very much in the minority by the late-nineteenth century. In court, precocious pregnancy shed doubts on a complainant's status as a child but was not in itself a marker of consent.

As pregnancy declined in importance as a sign of sexual consent, medical witnesses placed increasing weight on marks of resistance. In 1878, for example, writing on the subject of 'violation after puberty in virgins', medical jurist Francis Ogston stated that:

> There are some other physical marks of connection which, if found on the person, will lead pretty certainly to the conclusion that the connection has not been with the woman's consent. The party who has been violated will frequently present bruises on different parts of her body, such as she would not be likely to inflict on herself, but which are easily accounted for on the supposition that she has struggled against the embraces of her ravisher. These bruises are most likely to be found on the groins, thighs, wrists, or breasts, or about the throat.[30]

Between 1852 and 1894, Taylor's medical jurisprudence books similarly noted that girls over the age of sexual consent were 'considered to be capable of offering some resistance to the perpetration of the crime; and therefore in a true charge, we should expect to find not only marks of violence about the pudendum, but also injuries of greater or less extent upon the body and limbs'.[31] Such advice was put into practice regularly, as signs of resistance were relatively easy to identify and to interpret. In a Middlesex indecent assault trial from 1871, for an assault on a girl of unspecified age but sufficiently old to walk alone and go to a public house, the surgeon Thomas Warner Lacey testified that:

> I found that she had no marks upon the right thigh, but there were several scratches from the hip joint down to the knee on the left thigh. She said there were no other marks, there was no abrasion on the thigh but only red marks. They were recent marks, and might have been done within an hour. On her left cheek there was a red mark as if a blow had been struck. It might have been caused by her falling on the ground. The marks upon the thigh could not have been caused by any portion of her dress. The marks were irregular and in my judgement would indicate resistance.[32]

This testimony broadly followed the textbook advice, in which signs such as bodily 'abrasion' or 'scratches', and a facial 'red mark' indicated resistance. Occasionally medical witnesses refused to give their opinion about the origin of bruises, but the widespread use of phrasing such as 'violence or bruises' and 'there were no bruises indicating resistance' clearly conflated these issues.[33]

Witness testimony on the subject of consent and resistance focused on pubescent and adult women, but puberty was a flexible life stage in medical thought. In consequence, medical witnesses also made occasional reference to the presence or absence of signs of resistance in cases involving mature girls under the age of sexual consent. In October 1888, when a ten-year-old girl was allegedly assaulted in Devon, a local surgeon testified that:

> I examined her carefully. I found her a virgin. I examined her for bruises but I failed to find any or any scratches. I found nothing abnormal. She is a strong muscular girl who would not shew [sic] bruises unless some strong violence had been used.[34]

This complainant was comfortably under the age of sexual consent, within the felony clause. In law her resistance was irrelevant, but the medical practitioner's testimony that she was a 'strong muscular girl' implied a physical capacity to resist. His comments incorporated knowledge about the ambiguities of sexual development and created space for interpretation; due to her physical maturity, the girl's capacity to resist or consent was no longer automatically absent. Such testimony followed general medico-legal advice. William A. Guy and David Ferrier, in their 1881 *Principles of Forensic Medicine*, had noted that it was useful in cases of alleged rape to 'ascertain complainant's age, strength and state of health' as part of interpreting evidence.[35] Age was an important part of assessing physical signs, but the strength of a girl was not automatic. Her capacity to resist was assessed by a combination of textbook knowledge (age-based development) and case-by-case observation (physical strength). Once girls passed puberty, however, their strength was often assumed. In 1889 a general practitioner from Devon testified that 'I think it impossible for a man standing up to commit a rape on a girl who is conscious', drawing upon wider medical and social rape myths about the ability and responsibility for women to resist.[36]

Medical witnesses and authors showed little sympathy for post-pubescent female complainants. They actually increased the burden on women to show marks of resistance over the course of the Victorian and Edwardian periods, in contrast to the more sympathetic trend of case law. The 1910 edition of Taylor's *Principles and Practice* added the following:

> Bruises upon the arms particularly may be considered to be reasonable evidence of attempts at struggling, impressions of fingernails, too, would be suggestive. Strong corroborative evidence of a tale of struggle might be obtained from an examination of the accused for similar marks of bruises or scratches about the arms or face, and possible even about his penis, though this is much less likely.[37]

Taylor's wording raised the possibility that a woman's story of a struggle was no more than 'a tale'. He sought to limit wrongful conviction by encouraging medical witnesses to examine both the complainant and the prisoner for marks of resistance. This addition to the book indicated a growing link between resistance and violence, with 'physical proofs' of a struggle newly extended to the prisoner's body. Despite the shift away from physical models of consent in law, medical texts articulated increasing concern about false charges under the 1885 Criminal Law Amendment Act and highlighted women's capacity to self-inflict injuries. This change responded to a trend already evident in the courts, as seen in two Gloucestershire cases from 1896 and 1898. These cases were the first in which a medical witness testified about scratches visible on the prisoner, as well as on the complainant, although only to note their absence. In 1896 the medical witness stated that 'I did not find any marks or scratches on the prisoner except some signs of his nose having bled'.[38] In 1898 another local medical practitioner testified that 'I saw nothing on [the prisoner] to commit him with such a charge. There was no sign of recent scratches on his face'.[39] In this case, the prisoner himself cross-examined the medical practitioner to elicit such a response. Changes to medical textbooks acknowledged the greater burden of proof demanded within the courtroom, rather than necessarily driving this change.

Medical witnesses testified extensively on marks of resistance, especially once their remit expanded to include the prisoner's body, but were not permitted to give an opinion on whether these signs proved an absence of consent. As Guy and Ferrier noted in their medico-legal textbook in 1881, '[t]he sufficiency of the resistance, and the question of consent generally, are reserved for the jury'.[40] Medical witnesses could, however, give an opinion on the reasons for a lack of resistance if those reasons had a scientific basis. One such situation was when 'insensibility' was an alleged or suspected cause of a complainant's failure to resist. Insensibility was a broad concept, the use of which ranged from describing exhaustion to the effect of drugs or alcohol, but consistently explained why a complainant may have lost the strength to fight off an assault. In 1864, a Teignmouth surgeon testified that 'I found [the complainant] insensible and struggling in a fit ... Her state was such that as she came to herself her legs and arms trembled very much as if she had been exposed to long and repeated exertions, or from fright. The fits may have been occasioned from the like cause'.[41] This testimony explained why a healthy woman, 25 years of age, would fail to resist a man's advances. It helped to address the paradox of rape, which was both 'against the will' in law and deemed to be impossible if a healthy woman resisted. Medical testimony corroborated the complainant's claim of having struggled to the point of exhaustion.

Medical practitioners, counsel and judiciary interacted in the formulation of medical testimony and drew upon shared pools of medico-legal knowledge in relation to 'insensibility'. In consequence of the widespread influence of standard medical jurisprudence works such as Taylor's, legal counsel could pre-empt and encourage specific medical evidence. When one Middlesex complainant testified to falling insensible during an alleged sexual assault, *Reynolds's News* reported that:

> Robert Scott, F. R. C. S. [Fellow of the Royal College of Surgeons], deposed that he lived at Arundel-terrace, Barnebury-road. He was sent for to see Miss Luff, whom he found in Carroll's parlour ... [there were] no indications of violence on the girl ... (cross-examined) ... I do not know of a drug which would produce the immediate insensibility spoken of by the prosecutrix. Prussic acid will, but of course it is fatal. Aconite will produce insensibility. I know Taylor's *Jurisprudence* – it is a work of authority. Tincture of aconite will produce insensibility, but there are the intervals of screaming, and the patient remains for a long time under its influence. (Mr Sleigh) – Are not these the symptoms of the administration of aconite – flashing of fire from the eyes, giddiness in the head, and numbness of the extremities? (witness) – Yes, they are. (Mr Sleigh) – The very symptoms this girl has described.[42]

Another surgeon later conducted a second examination and found that bruises had developed on the complainant's body, which meant that proving a state of insensibility was no longer necessary to justify a lack of resistance. However, the process by which legal counsel elicited the first medical witness knowledge remains important. Taylor was right in noting 'the acquisition of much medico-legal knowledge by lawyers', not least from his own book.[43]

Although the mental aspects of consent were 'legal' rather than 'medical' questions, medical witnesses also spoke sporadically about mental (in)capacity. As Mark Jackson notes in his work on *The Borderland of Imbecility*, around the turn of the century increasing attention was paid to 'mild forms of mental deficiency' such as 'feeble-mindedness'. Mild forms of so-called 'imbecility' and mental 'incapacity' also came to influence sexual forensics, particularly after the 1885 Criminal Law Amendment Act protected such girls. As medical testimony had initially shaped case law on the subject of sexual instincts in 'imbecile' girls, the place of medical testimony on what Jackson calls 'mild ... mental deficiency' in court was secured. However, medical opinion always came second to that of the magistrate or judge. In an 1886 Middlesex case, which involved an alleged assault on a girl aged 17, the complainant's mother testified that she was 'afflicted in her intellect and not capable of taking care of herself. She doesn't know the meaning of anything wrong and is as harmless as a baby'.[44] The court asked a divisional surgeon of police to assess the complainant's mental competency. He stated that:

> I examined the girl Caroline Watson at 11 last night ... She appeared to be of very weak intellect. I do not think she is of capacity to understand an oath. (the doctor examined Caroline Watson at the court) I have now examined Caroline Watson. She appears more rational and I should say will understand the nature of an oath. (cross-examined) I should say she has sufficient capacity to consent to any act.[45]

This surgeon implicitly justified an expectation that Caroline Watson should have resisted, although found no signs of such resistance. Following this medical testimony, 'Caroline Watson was called and examined by [the] magistrate who concluded she was not of sufficient capacity to give evidence'.[46] The magistrate thus overrode the medical witness on the issue of the girl's capacity to understand an oath. The surgeon's inconsistency and the magistrate's dissent destabilised the surgeon's other claim about Watson's 'capacity to consent'. His testimony also indicated that mental capacity itself might be inconsistent, meaning that any assessment of a girl's mental capacity during a trial did not prove or disprove her capacity to consent at the time of the alleged offence.

Overall, despite showing an interest in the mental aspects of consent, medical witnesses had limited opportunity to speak on such matters and more typically focused on physical signs of resistance. The courts, in contrast, also took the psychological aspects of consent into consideration. This trend was not limited to the UK. As Stephen Robertson notes, in his work on rape in the United States from 1823 to 1930:

> Doctors came to the legal system with an understanding of rape as a physical struggle. As medical jurists gained more experience in the legal system and became more familiar with the details of legal definitions of rape, they found that legal discourse defined rape more broadly than they did. The knowledge and expertise claimed by doctors did not extend to all aspects of the legal definition of rape.[47]

In England, likewise, the courts used a broader definition of consent than medical witnesses. Medical literature showed awareness that 'against the will' might mean more than physical resistance, and occasionally touched upon the subject of mental capacity to consent, but emphasised that the social and legal dimensions of such questions were beyond the remit of sexual forensics. Medical testimony continued to focus on the mature female's responsibility to resist, even though the law began to shift some responsibility onto men to gain 'permission'.

Despite these differences between legal and medical models of consent, medical testimony about marks of resistance seemingly had a significant impact on trial outcomes. Before a case even reached trial, magistrates and grand juries often dismissed cases on the basis of insufficient resistance. As a result of this filtering process, there was evidence of bodily violence in 40 per cent of pre-trial statements (across all courts) involving complainants over the age of 16 and medical testimony; this statistic was almost double the equivalent for signs of genital violence. Marks of resistance were not officially required by statute because the type and degree of evidence required to prove that a sexual act was 'against the will' was a matter for the court's discretion.[48] However, newspapers often reported that cases were dismissed when magistrates believed that the complainant had shown insufficient resistance. In a case of suspected rape on a young Gloucestershire servant girl, who was over the age of consent in July 1885, *The Bristol Mercury and Daily Post* reported that:

Dr Andrew Currie deposed that there were no marks of violence upon the prosecutrix except that her arms were scratched somewhat. In his opinion prosecutrix could not have resisted very much, or there would have been greater evidence of injury. The bench dismissed the case.[49]

This case was not dismissed because of a lack of any signs of resistance, but because the court deemed the degree of resistance to be insufficient. Medical evidence was seemingly decisive in this case, although the subjects of consent and resistance were not purely scientific matters. *The Bristol Mercury and Daily Post* also reported in 1891 that a Somerset magistrate dismissed a case of suspected rape 'without hearing the medical or other witnesses' after a 17-year-old complainant admitted that she did not call for assistance or immediately complain of the alleged assault.[50] This girl's own testimony, rather than that of a medical practitioner, proved her lack of resistance. Magistrates' decisions to dismiss cases or commit them for trial turned on very similar issues, irrespective of whether a medical man provided the evidence; medicine added weight to such concerns, but did not create them.

Medical witnesses guided decisions about consent and resistance in a range of implicit ways. Magistrates drew not only upon evidence about physical harm and resistance, but also upon medical testimony about a complainant's general character. As the previous case analysis showed, evidence of previous unchastity increased the burden of evidence for resistance due to a general expectation that unchaste girls were more likely to give consent. When a 'decently-attired' man was accused of raping a 14-year-old girl in 1858, a surgeon found that the girl had had sexual intercourse within 24 hours but found no signs of violence or recent loss of virginity. The magistrate therefore concluded that:

> [F]rom what he could ascertain from the surgeon's evidence it was not the first thing of the sort the girl had been corrupted with; and although the prisoner had done wrong, still he could not say a rape had been committed, as the girl had said nothing about the matter, had not called out, and did not scratch his face. – The prisoner was then discharged.[51]

The difference between the prisoner's apparent respectability and the working-class girl's previous unchastity was evident in this newspaper report. The girl's sexual history exacerbated her apparent lack of resistance, defined here in terms of very specific signs such as calling out and scratching the prisoner's face.

Reports of trials from the Middlesex Sessions and south-west Quarter Sessions indicate that juries, like medical witnesses, drew connections between physical resistance and consent. Although the question of consent commonly excluded children, it was legally relevant in indecent assault cases involving male or female complainants of any age before 1880 and over the age of 13 after 1880. In 1857 *The Times* reported on a case of alleged indecent assault against an eight-year-old girl in which 'as it appeared the child did not resist [the prisoner] was acquitted on the point of law as to consent'.[52] Carolyn Conley notes similar cases in her work on crime in Victorian Kent, in which she argues that juries often conceptualised consent in terms of physical resistance rather than as a matter linked to 'consent of will', fear or influence. She notes that juries focused on physical resistance in incest cases despite the apparent mitigation of the unequal father-daughter power relationship, citing two cases in which prisoners were acquitted because their teenage daughters showed 'insufficient resistance' to an alleged assault.[53] Notably, in these cases resistance was deemed 'insufficient' rather than entirely absent.

The tendency for juries to focus on consent as a physical struggle, sometimes irrespective of a complainant's age or relationship to the prisoner, meant that verdicts often corresponded to medical findings about bodily violence. For all 'true bills' without guilty pleas, juries convicted in 70 per cent of cases in which medical witnesses found marks of violence on a complainant's body and a notably lower 57 per cent of cases in which medical practitioners explicitly noted the absence of such marks. Because juries were often directed on the subject of consent, it is possible that they felt obliged to adhere to the official law on the matter even in cases involving young girls and boys. This possibility is supported by a Middlesex case involving an alleged indecent assault on a nine-year-old boy in 1857, in which a prisoner was acquitted on the basis of the boy's apparent consent. In this case, according to newspapers, the verdict was passed 'with regret' after the assistant judge 'left the case to the jury upon the question of consent'.[54] However, such direction allowed juries some discretion to define consent and they remained wedded to links between 'real rape' and physical resistance, even after changes to case law.

Medical testimony could also contribute to a jury's perception of whether a female or, less commonly, male complainant was 'insensible' in order to justify a lack of physical resistance. In one Middlesex case involving an 11-year-old male complainant, the medical witness testified that '[h]e was tossing about on the bed much excited. I thought him drunk but found no perfume of drink in his breath. He made a statement to me. He said he became insensible'.[55] This testimony, which described the boy as appearing drunk but having no signs of alcohol intake, implicitly corroborated the boy's claim to have been drugged. *Reynold's Newspaper* reported that the prisoner:

> [W]as indicted for attempting feloniously to assault a boy eleven years of age, after administering to him chloroform, or some other noxious drug. Medical evidence having been heard, the jury found the prisoner 'Guilty,' under the third count of the indictment, of an indecent assault.[56]

This report indicated a connection between the jury's verdict and the medical testimony. However, newspaper reports also indicate that lay witness testimony on 'insensibility' influenced juries in much the same way as similar medical evidence. In a Devon case of suspected indecent assault on a young adult female, the local newspaper reported that '[a]fter serving the prosecutrix in a very brutal manner – so much so that she was insensible when found – the prisoner left her, and made his way off ... The prisoner was found guilty'.[57] In this case proof of insensibility came not from a medical practitioner, but from the witness who found the girl on the floor after the alleged assault.[58] The power of insensibility as a concept lay not in its medical nature, but in its social significance. It was a necessary part of explaining why a complainant might fail to resist a rape and connected with wider social models of frail victimhood, which applied both to young boys and adult women.

Judges took a broader view of the meaning of consent and, in consequence, did not draw upon medical testimony to the same extent as jurors. Judges, as in case law, seemingly conceptualised consent in more than physical terms. Few records survive of judges' decision-making processes but, when combined with case law decisions, it seems significant that judges passed heavier sentences in cases involving possible 'submission' rather than 'consent'. They passed high sentences in cases of 'carnal knowledge' that involved incest, for example, irrespective of the degree of resistance that a complainant had demonstrated. Prisoners who committed incestuous assaults on young female family members were given some of the highest sentences: across all cases the mean average sentence for direct blood relatives was 84 per cent of the maximum by law, and the same statistic for an uncle or a relative by marriage was 70 per cent of the maximum sentence for law. *The Times* reported on a Middlesex case from 1875 in which:

> James Seagrave, 58, was indicted for attempting to ravish and for assaulting with intent his step-daughter, a girl under the age of 13 years ... Mr. Edlin said it was as gross an outrage, considering the position the prisoner occupied towards the child, as had ever come before him, and sentenced him to two years' imprisonment, with hard labour.[59]

These responses to incest cases may have been due to broader anxieties about the 'unnatural' nature of incestuous relationships between blood- and step-relations, rather than to ideas about consent. However, the impact of incestuous relations on sentences also tied in with case law decisions in which familial relationships were noted to cause 'mere submission'. These models of consent differed from the emphasis on pure physicality found in much medical testimony and in jury verdicts.

Consent was undoubtedly central to many trials in both legal and moral terms, particularly those involving female complainants, but its definition was not clear-cut. As judges and jurors drew upon slightly different understandings of consent, the significance that they placed on medical testimony about bodily resistance was not consistent. Judges understood consent to be a multifaceted issue and jurors placed greater emphasis on the physical struggle, therefore the latter group were particularly receptive to medical evidence on bodily violence. Although judges' and juries' conceptualisations of consent might have merged in cases with directed verdicts, juries also had discretion in the judicial process. Magistrates passed cases forward to trial that fitted jurors' expectations about sexual crime, often reinforced by media reports that constructed rape as an inherently violent offence, thereby strengthening these stereotypes. Medical testimony was received and interpreted in light of judges' and jurors' pre-existing ideas, but may also have served to reinforce 'real rape' stereotypes in which resistance was a matter of bodily violence.

Criminologist Gethin Rees notes that ' "real rape" typically involves an unsuspecting woman being attacked by a stranger, in an outdoor location at night, with the stranger employing force or a threat of force (with the use of a weapon), and the victim offering active resistance'.[60] The centrality of 'force' to Rees' framework is key to understanding not only current-day but also historic attitudes to sexual crime. Magistrates often dismissed cases before trial if they did not fulfil the stereotype of violent rape against adult women, as violence (interpreted as resistance) was a crucial part of proving the absence of consent. Medical witnesses played a key role in consolidating such models of 'real rape'. Despite the continuity evident in Rees' definition of 'real rape', these Victorian and Edwardian models of sexual resistance also differed in some ways from those found today. With late twentieth-century feminism, when rape was redefined as a crime of power relations rather than of sexual desire, there emerged a notion of males prone to violence for its own sake rather than as a response to the actions of a victim. However, individual judges continued to draw upon nineteenth-century language and conflate force with resistance in the late decades of the twentieth century. Judge David Wild, for example, a judge at the Cambridge Crown Court, summed up a trial in 1982 as follows:

> Women who say no do not always mean no. It is not just a question of saying no, it is a question of how she says it, how she shows and makes it clear. If she doesn't want it she only has to keep her legs shut and she would not get it without force and there would be marks of force being used.[61]

Wild did not represent the views of all judges and the extensive citation of this statement makes evident its atypical nature by the late-twentieth century. In 2002 Hilaire Barnet described this statement as 'extraordinary' and noted that 'perhaps the sensitivity of judges has improved' in response to widespread criticism of such comments.[62] Despite its exceptional nature, the case shows that, almost a century after some judges sought a more nuanced conception of consent, other judges still clung to rape myths that placed heavy weighting upon medical evidence of resistance. It was only in 2000 and 2003 respectively that the law recognised 'abuse of a position of trust' and 'sexual grooming' as factors contributing to sexual consent. At the demand of certain courts, sexual forensics reinforced models of violent 'real rape' long after those models had been socially and legally contested.

Case Analysis: Consent, 1877

The Case: This case is unusual because it is one of two complaints brought to court by the same girl, labelled an 'imbecile'. It provides a unique opportunity to consider how age, mental capacity, sexual development and the law on consent interacted to shape medical and judicial attitudes to consent and resistance (or a lack thereof). It shows the limits of imbecility as a justification for the absence of resistance and has general significance for issues raised in the associated chapter, such as the moral dimensions and physical construction of consent/resistance.

The Prisoners: Walter Trass, aged 50, Labourer.
The Complainant: Rosina Symes, aged 13.
The Complaint: Attempted carnal knowledge (consent a defence).
The Pre-Trial Statement: Depositions taken 22 September 1877, Temple Cloud Police Court.[63]

William Symes: I am a gardener and reside at Chew Magna in this County. On Friday last the 7th of September ... between 12 and one o'clock in the day I was at work in some ground near Chew Magna. Mrs Catch called me. I ran to her by a gate and she said she had seen something that ought not to be with my daughter. She said she had just lost sight of her. I ran along the road and turned up a green lane where there was a field gate, and under the hedge there I saw the prisoner upon my daughter on the ground. His clothes were down and my daughter's were up. They were on the grass together. When I spoke to him he rose up. My daughter's person was exposed, they appeared to have been in connexion together, but on his rising his shirt fell down. He said several times 'don't say anything about it'. I said 'I shall do no such thing' and called to my master's coachman who was working with me in the field ... I placed the prisoner in his hold until I fetched a policeman. My daughter was 13 years old last February. She is weak of intellect, part an idiot, and not fit to give evidence in the case. The distance from where I found the prisoner in the field with my daughter is about 300 yards from my house. The prisoner is a stranger in the neighbourhood. I never saw him before. (cross-examined by prisoner) You was down right upon my daughter, you appeared to be having connexion with her ... I did speak to you ...You several times asked me not to say anything about it. (re-examined by the bench) When I saw my daughter and the prisoner there was no struggle going on between them. I did not see my daughter make any resistance. I heard no words and no cries. My daughter did not speak to me. I said to her 'you look sharp get home'. When I first caught sight of the prisoner and my daughter I was at the field gate and they were lying on the grass within two or three yards of the gate, their heads towards the hedge.

Martha Patch: I am the wife of Charles Patch and reside at Chewstoke in this County. On Friday the 7th of September last I had been to Chew Magna, and on the High Road on my return towards home at about 12 or 1 o'clock [I saw] a man, whom I identify as the prisoner. I stood in the road and saw the girl pass the man, look at him, keeping her eyes upon him, and she then went back, and sat down by the side of him on the bank. I then lingered in the road and saw the man and girl walking side by side. At the top of the hill directly after I lost sight of them, I then saw the witness Mr Goodland. I said nothing to him. I then went to the girl's father who was working ... and told him that something was going on. It was about 100 yards from the place to where I saw her father Mr Symes. I told him that I thought something was improper going on with his daughter and some strange man, upon which he went as fast as he could in the direction ... (cross-examined by prisoner) You were either sitting or lying against the bank. I did not see you do anything. I saw Mr Goodland pass. I did see you. I lingered purposely.

Job Goodland: I am a butcher and reside at Chew Magna in this County. On Friday the 7th of September last. I was coming from Chewstoke to Chew Magna. I remember meeting the witness Mrs Patch, she was walking in the direction of Chewstoke. Then I met Rosina Symes directly after passing Mrs Patch. She was walking in the direction of Chewstoke. I then saw the prisoner pick up a reap hook from a bank in the road. He was about 10 yards from Rosina Symes. He then walked in the direction of Chewstoke after her. I saw him follow her. He was close behind her. There is a green lane ... out of the road to a field. I know the lane, its length is from 20 to 25 yards. I had just passed this lane when Rosina Symes came up. I then drove on – I heard no words or any crying out. The lane leads to a field. There is a gate there.

George Sydenham: I am a Police Constable stationed at Chew Magna in this County. On Friday the 7th of September last from what I was told I took the prisoner into custody about 1 o'clock in the day. I charged him with attempting to carnally know one Rosina Symes. He was about to make a statement but I cautioned him. Afterwards he said 'I was on the road from Chew Magna to Chewstoke. I sat on the bank there. This girl', he said, 'came and sat down by the side of me and caught hold of me by the arm. Afterwards we went on the road a little way together. I told her that I wanted to go over this gate to do a job for myself and ... the girl came over the gate and sat down on the grass'. He said 'I didn't touch her'. Sometime after this on the way to the station he made another statement to the following effect: 'I don't know what they can do with me. I never put anything into the girl nor injured her in any way but I don't know what I should have done if the father had not come up so soon'. I heard prisoner repeatedly ask the prosecutor to forgive him and not press the charge against him. He was sober at the time. He is a stranger.

James Laver: I am a Coachman ... and was in a field with the prosecutor ... on Friday the 7th September last. I recollect some person calling him to the public road, I saw him then run along the road. I went down to the road, as I thought something was the matter. I looked over the wall, I heard prosecutor calling. I got over the wall and went down the road, to the end of the green lane. I found the prosecutor with his daughter and the prisoner at the end of the Lane. Prosecutor said 'this man has had to do with my girl'. Prisoner said 'I have done the girl no harm, let's make it up'. Prosecutor said he should do no such thing and he asked me to look to him whilst he fetched the policeman, which I did. Prisoner wished me let him go. After waiting some time I proposed to walk to the policeman's, which we both did, but he was not at home. Afterward I delivered him to the witness Police Constable Sydenham.

Charles Howell Collins: I am a Surgeon residing at Chew Magna. Emma Symes, the mother of Rosina Symes brought her daughter to my surgery at Chew Magna in the middle of the day of the 7th of September for me to examine her, stating that she had had a criminal assault committed upon her. It was about 2 o'clock when I examined her. I found a little redness about the orifice of the vagina and a slight tenderness as she complained that my examination hurt her. I found a little mucus at the orifice of the vagina. I am not able to say from my examination that any connexion had taken place recently. The evidence of my examination satisfied me that she was not a virgin. No injuries appeared upon her person. I consider her not capable of giving evidence being quite an imbecile. Yesterday I saw her with several men. They complained to me that they could not get rid of her. I ordered her home. I called her brother who was some distance off to come and take her home. I don't consider her capable of consenting to or dissenting from an attempt upon her person.

Walter Trass: When I went into the field I went and done a job for myself. I came back to the gate and pulled my boot and stocking off and just as I got my stocking off the girl came over the gate and sat down about 2 yards from me. I asked her where she was going and waving with her hand said 'I live out here'. I put on my stocking and my boot and said no more to her then. Whilst I was lacing up my boot she caught hold of me by the tie of my trousers. She had her clothes up to her knees and in the meantime her father came to the gate and said to me 'what are you doing with my daughter here?'. I said 'nothing at all'. That is all I have to say.

The Verdict: No Bill.

Significance for Sexual Forensics: The significance of this case becomes evident when it is compared with another case from the previous year, in which Rosina Symes also complained of an assault. In 1876, Symes was aged 12 and was allegedly the victim of an assault involving penetrative sex. Although recorded as 'indecent assault', for which the girl's consent would have been a defence at this time, notes on the pre-trial statement cited the 1875 law on sexual consent: '38 + 39 Vic. c. 94. Abusing + c. a girl above 12 + under 13. With or without consent. Misdemeanour. Not exceeding 2 years impris[onmen]t'.[64] According to her mother, Rosina Symes was 'incapable of giving evidence, from imbecility and is half an idiot owing to suffering from fits', but a potential medical witness was not called to corroborate these claims in 1876.[65] The prisoner admitted the offence and claimed that the girl had initiated the sexual contact, but – perhaps due to a misapplication of sexual consent law to an indecent assault case – the court did not probe claims about her imbecility, consent or resistance. The jury reached an easy 'guilty' verdict.

In 1877, in the case transcribed above, Rosina Symes came before the court again as the apparent victim of a second assault by a different man. Then aged 13, she was over the age of consent and needed to show resistance. Again, the relatives of Rosina Symes raised the issue of her mental capacity. In this case, which differed from the first in the specific attention given to the matter of resistance, her father noted the absence of a struggle. The surgeon called was the same man consulted, but not called to testify, the previous year; as the girl's consent became of interest to the court, so did his testimony. Even though 'imbecility' was not yet an official part of sexual consent law, it was deemed relevant to the question of consent. The surgeon indicated that Symes was not a virgin before the alleged assault in 1877 and that she had a tendency to pursue males, both of which undermined the complainant's case.

The girl's ostensible unchastity may have been a result of the previous assault in 1876, rather than of consensual sexual activity. Although the medical witness made no explicit reference to this case, the same practitioner had been consulted in both and was aware of the previous allegation. The previous year, however, this medical practitioner had not been given the opportunity to testify. It was therefore left unclear whether or not the girl was a virgin before the first offence. The cause of Symes' lack of virginity was, however, less important than its implications for the question of consent. In contemporary thought a girl who had 'fallen', whether consensually or in consequence of an assault, was more likely to be 'polluted' and promiscuous. Medical testimony about a lack of virginity at the age of 13, irrespective of its original cause, raised questions about the girl's status as innocent child. This practitioner also drew upon his role as a member of the local community to testify to the girl's general conduct, when he stated that '[y]esterday I saw her with several men. They complained to me that they could not get rid of her'. He provided both social and scientific evidence to support the prisoner's claim that the girl pursued him.

The medical testimony in this case addressed a number of moral and legal issues. Although medical testimony about the girl's sexual history and behaviour raised doubts about her innocence, the surgeon's testimony about mental capacity provided a scientific explanation for Symes' ostensibly immoral behaviour and supported the prosecution. The medical witness testified that the girl was an 'imbecile', which explained her apparently forward behaviour and lack of resistance. Medical witnesses were permitted to speak about the capacity to consent in such circumstances, as imbecility was a medical diagnosis rather than a legal category. The surgeon gave his opinion that she was not 'capable of consenting to or dissenting from an attempt upon her person'. In theory, this testimony cancelled out the other testimony about her sexual history and character. Previous unchastity or abuse indicated that the girl was more likely to consent, but her 'imbecility' provided a medical explanation for her lack of resistance. Consent and resistance were mental as well as physical in such circumstances, differing from most medical testimony about 'normal' pubescent and adult females.

Despite the legal implications of this medical testimony, which removed the responsibility to resist, the grand jury declared a 'no bill'. The grand jury may have been hesitant about convicting a second man in the light of Symes' apparent habit of pursuing men. They may also have considered that she showed signs of 'animal instinct', which had constituted a form of consent in case law since the 1860s. In the 1870s many juries still believed that a prisoner was only obliged to respond to dissent, rather than to elicit active consent. That the prisoner, a stranger to the area, knowingly took advantage of her imbecility would have been impossible to prove. In other cases in which the judge's direction is recorded, including the few recorded Old Bailey rape trials, this issue of a prisoner's knowledge was an evident concern. In 1906, a judge in an Old Bailey rape case 'left two questions to the jury: – (1) Was the woman an imbecile? (2) Did prisoner know it?'.[66] The reasoning of the grand jury in the Symes case is not recorded in the same way, but the case had a direct witness and did not lack medical evidence that a sexual act had taken place. The lack of *prima facie* evidence therefore must have related to a lack of evidence that the crime was knowingly 'against the will'. Medical testimony spoke only to the first question ('was [she] an imbecile?') but could not prove that the prisoner knew of this imbecility, which was required to confirm his guilt. Legal issues also did not necessarily outweigh moral concerns, shared with medical witnesses, about the girl's uncontrolled sexuality and apparent precocity.

Notes

1. Lawson Tait, 'An Analysis of the Evidence in Seventy Consecutive Cases of Charges made under the New Criminal Law Amendment Act', *Provincial Medical Journal*, 1 May 1894, 226–235, p. 227.
2. 'Criminal Jurisprudence – Rape', *The Medical Press*, 23 July 1890, 87–89, p. 87.
3. On this myth see Joanna Bourke, 'Sexual Violation and Trauma in Perspective', *Arbor-Ciencia Pensamiento Y Cultura* 743 (2010), 407–16, p. 410.
4. Gurney Williams, 'Rape in Children and in Young Girls', *International Clinics* 23 (1913), 245–67.
5. For further details on male lust, self-control and ideas about the sexual offender, see Chapter 6.
6. Lydia Morris, *Dangerous Classes: The Underclass and Social Citizenship* (London, New York: Routledge, 1994), p. 2.
7. Alfred Swaine Taylor, *Medical Jurisprudence*, 4th edn (London: J. & A. Churchill, 1852 [1844]), p. 587.
8. Richard Burn, *The Justice of the Peace and Parish Officer and Parish Officer*, 30th edn, vol. 1 (London: H. Sweet, Maxwell & Son and Stevens & Sons, 1869), p. 309–10; David Taylor, *Hooligans, Harlots, and Hangmen: Crime and Punishment in Victorian Britain* (Santa Barbara, CA: Praeger, 2010), p. 119.
9. Burn, *Justice of the Peace*, p. 310.
10. The Queen v Lock (1872) LR 2 CCR 10; see also Regina v McGavaran (1852) CCLC 6.
11. Steve Hedley, 'Brett, William Baliol, First Viscount Esher (1815–1899)', *Oxford Dictionary of National Biography* (Oxford: Oxford University Press, 2004) <http://www.oxforddnb.com/view/article/3350> (accessed 23 July 2015).
12. Thomas Dixon, 'The Tears of Mr Justice Willes', *Journal of Victorian Culture* 17 (2012), 1–23, p. 2.
13. Regina v Woodhurst (1870) CCLC 12.
14. Regina v Rudland (1865) 4 F and F 495.
15. Elizabeth Kolsky, ' "The Body Evidencing the Crime": Rape on Trial in Colonial India, 1860–1947', *Gender & History* 22 (2010), 109–30.
16. R v Fletcher (1866) LR 1 CCR 39.
17. R v Fletcher (1866).
18. R v Fletcher (1866).
19. The prisoner was found guilty, but it remains significant that the issue of 'animal instincts' was cited as part of the judge's deliberations; R v Barratt (1873) LR 2 CCR 81.
20. It was not thought to be 'real' consent if women allowed a fraudulent medical practitioner to conduct an intimate examination or if they consented, in the dark, to sexual intercourse with a man thought to be their husband. The Queen v Flattery (1876–77) LR 2 QBD; The Queen v Barrow (1865–72) LR 1 CCR 156.
21. R v Dimes (1912) 7 Cr App R 43.

22. R v Dimes.
23. Throughout the editions of Taylor's *Manual*... and *Principles and Practice* he claimed that women could resist, setting aside only 'infants, idiots, lunatics, and weak and delicate females'. Only one of these categories was deemed a natural product of age, until Taylor added older women in 1910. For example see Taylor, *Medical Jurisprudence*, 4th edn, p. 587; Alfred Swaine Taylor, *The Principles and Practice of Medical Jurisprudence*, ed. Thomas Stevenson, 6th edn, vol. 2 (London: J. & A. Churchill, 1910 [1865]), p. 112.
24. Charles Roberts, 'The Physical Maturity of Women', *The Lancet*, 25 July 1885, 149–50, p. 149.
25. Taylor, *Principles and Practice*, 6th edn, p. 130.
26. Taylor, *Principles and Practice*, 6th edn, p. 130.
27. Francis Ogston's 1878 *Lectures on Medical Jurisprudence* described one nine-year-old complainant as 'precocious' because her 'familiarity with the usual details connected with sexual intercourse showed that she was no stranger to the subject'; Francis Ogston, *Lectures on Medical Jurisprudence* (London: J. & A. Churchill, 1878), p. 93.
28. Bourke, 'Sexual Violation', p. 409.
29. J. Matthews Duncan, 'Gulstonian Lectures on Sterility in Woman', *The Lancet*, 31 March 1883, 517–29, p. 529.
30. Ogston, *Lectures on Medical Jurisprudence*, p. 116–17.
31. Taylor, *Medical Jurisprudence*, 4th edn, p. 582. This quote remained unchanged throughout the editions of this book.
32. London, London Metropolitan Archives (LMA), Pre-Trial Statements, Robert Feutom and Henry Oakshott tried at the Middlesex Sessions on 18 October 1871 for indecent assault, MJ/SP/E/1871/020.
33. London, LMA, Pre-Trial Statements, Henry Emeny tried at the Middlesex Sessions on 28 August 1878 for indecent assault, MJ/SP/E/1878/027; London, LMA, Pre-Trial Statements, Richard Belcher tried at the Middlesex Sessions on 7 April 1869 for indecent assault, MJ/SP/E/1869/007.
34. Exeter, Devon Record Office (DRO), Pre-Trial Statements, William Helmore tried at the Devon Quarter Sessions on 17 October 1888 for indecent assault, QS/B/1888/Michaelmas.
35. William A. Guy and David Ferrier, *Principles of Forensic Medicine*, 5th edn (London: H. Renshaw, 1881 [1844]), p. 71.
36. Exeter, DRO, Pre-Trial Statements, Frederick Trembath tried at the Devon Quarter Sessions on 16 October 1889 for indecent assault, QS/2/1889/Michaelmas.
37. Taylor, *Principles and Practice*, 6th edn, p. 130.
38. Gloucester, Gloucestershire Archives (GA), Pre-Trial Statements, George Henry Kirby tried at the Gloucestershire Quarter Sessions on 8 April 1896 for assault with intent, Q/SD/2/1896.
39. Gloucester, GA, Pre-Trial Statements, Thomas Cobb tried at the Gloucestershire Quarter Sessions on 19 October 1898 for assault with intent, Q/SD/2/1898.
40. Guy and Ferrier, *Principles of Forensic Medicine*, 5th edn, p. 70.
41. Exeter, DRO, Pre-Trial Statements, William Rhodes tried at the Devon Quarter Sessions on 22 November 1864 for assault with intent, QS/B/1864/November.

42. 'Shocking Case of Violation', *Reynolds's Newspaper*, 14 September 1856, 14, p. 14. Emphasis added.
43. Taylor, *Principles and Practice*, vol. 1 (London: J. & A. Churchill, 1865), p. 19.
44. London, LMA, Pre-Trial Statements, Henry Martin tried at the Middlesex Sessions on 10 August 1886 for indecent assault, MJ/SP/E/1886/040.
45. London, LMA, Pre-Trial Statements, Martin, MJ/SP/E/1886/040.
46. London, LMA, Pre-Trial Statements, Martin, MJ/SP/E/1886/040.
47. Stephen Robertson, 'Signs, Marks, and Private Parts: Doctors, Legal Discourses, and Evidence of Rape in the United States, 1823–1930', *Journal of the History of Sexuality* 8 (1998), 345–88, p. 348.
48. Carolyn A. Conley, 'Rape and Justice in Victorian England', *Victorian Studies* 29 (1986), 519–36, p. p. 520.
49. 'District News: Alleged Rape', *The Bristol Mercury and Daily Post*, 31 July 1885, 7, p. 7.
50. 'Alleged Rape at Kilmersdon', *The Bristol Mercury and Daily Post*, 10 October 1891, 6, p. 6.
51. 'Atrocious Cases', *Lloyd's Weekly Newspaper*, 25 July 1858, 9, p. 9.
52. 'Middlesex Sessions', *The Times*, 11 February 1857, 12, p. 12.
53. Carolyn A. Conley, *The Unwritten Law: Criminal Justice in Victorian Kent* (Oxford: Oxford University Press, 1991), p. 121.
54. 'Middlesex Sessions', *Lloyd's Weekly Newspaper*, 17 May 1857, 4, p. 4.
55. London, LMA, Pre-Trial Statements, Stephen Casaya tried at the Middlesex Sessions on 25 October 1870 for indecent assault on a male, MJ/SP/E/1870/023.
56. 'Middlesex Sessions', *Reynolds's Newspaper*, 30 October 1870, 6, p. 6.
57. 'Devon Epiphany Sessions', *Trewman's Exeter Flying Post*, 15 January 1857, 3, p. 3.
58. Exeter, DRO, Pre-Trial Statements, William Ponsford tried at the Devon Quarter Sessions on 7 January 1857 for assault with intent, QS/B/1857/Epiphany.
59. 'Middlesex Sessions', *The Times*, 28 August 1875, 9, p. 9.
60. Rees also notes that '[a]ttrition rate studies have discovered that attacks similar to that described by the "real rape" discourse are actually quite rare'; Gethin Rees, ' "It is Not for Me to Say Whether Consent Was Given or Not": Forensic Medical Examiners' Construction of "Neutral Reports" in Rape Cases', *Social & Legal Studies* 19 (2010), 371–86, p. 372.
61. Judge Wild cited in Carol Smart, *Feminism and the Power of Law* (London: Routledge, 2002), p. 35.
62. Hilaire Barnett, *Britain Unwrapped: Government and Constitution Explained* (London: Penguin, 2002), p. 14.
63. Taunton, Somerset Heritage Centre (SHC), Pre-Trial Statements, Walter Trass not tried (no bill) at the Somerset Quarter Sessions in October 1877 for attempted rape, Q/SR/709. Full transcript, except for occasional sentences with illegible words. Some punctuation added for clarity.
64. Taunton, SHC, Pre-Trial Statements, Benjamin Weaver tried at the Somerset Quarter Sessions on 19 October 1876 for indecent assault, Q/SR/705.
65. Taunton, SHC, Pre-Trial Statements, Weaver, Q/SR/705.
66. *Old Bailey Proceedings Online* (www.oldbaileyonline.org, version 7.0), trial of Frederick Briden in November 1906, t19061119—53.

5
Emotions: Medicine and the Mind

In 1856 London's *Daily News* reported a case of alleged indecent assault on 13-year-old Frances Young in which, when testifying at the police court, 'the complainant ... cried bitterly while giving her evidence'.[1] The complainant's conduct on the stand was seemingly as important as her testimony. The *Daily News* painted the girl in a sympathetic light by reporting on her 'bitter' tears, her 'modest' appearance and her mother's testimony about the girl's 'dreadful agitation' after the alleged offence.[2] Frances Young's appropriately feminine behaviour informed the magistrate's 'imperative duty to submit such a flagrant case for the decision of a jury', even though the prisoner was a 'respectably-dressed' man.[3] The case indicates that trials for sexual offences could hinge as much on the conduct, emotions and behaviour of a witness as on physical harm or witness testimony.

Although court depositions did not include details of witness conduct or appearance on the stand, they did record witness testimony about a complainant's emotional state in the immediate aftermath of an alleged offence. Such testimony reinforced expectations that females and children were close to nature and should demonstrate a significant emotional response to an offence. The medical and judicial interest in emotions created yet another a burden upon complainants; the 'real' victim in Victorian and Edwardian eyes was not only previously chaste and visibly unwilling, but also had the correct type and degree of emotional response without being over-emotional or hysterical. A true victim would be like Frances Young: generally modest; greatly agitated immediately after an offence; and emotionally delicate when made to recount the details of a crime.

There were social, legal and medical expectations that a young or female victim of sexual crime should demonstrate heightened emotions in the aftermath of a sexual crime. These expectations were never formalised in law or medicine, but were an important part of courtroom scripts. Despite the emotional dimensions of the courtroom, few scholars have applied the recent 'emotional turn' of historical research to medico-legal matters.[4] Historians of sexual crime have generally been more interested in the subject of trauma than emotions. Joanna Bourke is one of the few scholars to have paid attention to the emotional repercussions of rape in the nineteenth century, but focuses on general social ideas about emotions rather than how these operated in the judicial context. Bourke finds that the emotional repercussions of rape received little attention before consciousness-raising feminist movements; as late as 1957, she notes, a study of Sexual Offences 'devoted only a couple of sentences to the emotional responses of rape victims' in its 548 pages.[5] This context differed from the courtroom, where gendered emotional states and signs were central themes of witness testimony. The interpretation of bodily signs and emotional states – considered below in turn – were grounded not in research about body-mind relationships, but in assumptions about the emotional delicacy of particular social groups.

Peter N. Stearns and Carol Z. Stearns distinguish between the experience of emotions and 'emotionology': 'the attitudes or standards that a society, or a definable group within a society, maintains towards basic emotions and their appropriate expression'.[6] The experience and social construction of emotions are inseparable as, in line with Judith Butler's argument that gender identities are constructed through 'a stylized repetition of acts', emotional 'acts' were as much a part of gender identity as of performance.[7] However, any study of emotions in court is primarily a study of 'emotionology'. Descriptions and performances of emotional states were heavily mediated within trials, as part of general courtroom scripts. The emotions were a relatively small part of these scripts, compared with other evidence such as direct witness testimony or medical testimony on injury. However, it is significant that defence teams focused on perceived under- or over-emotionality in the wake of an alleged offence; these processes served to reinforce stereotypes of the emotional female victim and of the hysterical, lying women.

It is easy to overlook the role of the emotions, or 'emotionology', in sexual forensics because books of medical jurisprudence focused on the physical and moral repercussions of crimes. They paid little attention to the mind, beyond noting a possible link between fear and the absence of resistance.[8] The lack of connection between trauma and sexual crime was not limited to Britain, but reflected a general trend within European thought. Links between 'mental trauma' and childhood sexual experience can be traced back to 1890s European works by Sigmund Freud and Joseph Breuer, in the form of repression, and Pierre Janet, in the form of dissociation. However, Freud officially retracted his seduction theory in 1897 and few other professionals challenged his stance on the subject.[9] While other models of trauma and traumatic memory gained social and scientific acceptance in the early-twentieth century, particularly war trauma, it was not until anti-rape campaigns of the late-twentieth century that trauma and (child) sexual abuse became closely connected. However, Victorian and Edwardian mainstream medical literature did pay attention to some of the emotional and 'nervous' dimensions of sexual crime. Short-term emotional distress, although not yet conceptualised as chronic psychological trauma, constituted important evidence at trial. In the absence of any widespread belief in a link between psychological trauma and sexual crime in the late-nineteenth and early-twentieth centuries, the emotions were a focus of questions about how a sexual offence affected the mind.

There was extensive research on the emotions in the Victorian period. In the mid-nineteenth century, emotions were understood to be outward indicators of feelings. William Fleming's *Manual of Moral Philosophy* from 1867, for example, worked from the definition of emotions as 'feelings which, while they do not spring from the body, do yet manifest their existence and character by their peculiar influence upon the body'.[10] Over the course of the late-nineteenth century, as medical professionals also took an interest in the emotions, these models of the body-mind relationship were revised. In 1884 the American psychologist William James presented an influential alternative argument in 'What is an Emotion?' that the 'feelings' or 'emotions' resulted from a perception of instinctive bodily changes after an event rather than *vice versa*.[11] In this framework, physiological and behavioural changes were not manifestations of the emotions but the *cause* of emotions. James' ideas had roots in Darwinian ideas about bodily instincts, which required no cognition, in response to perceived threats.[12] Thomas Dixon cites this 1884 article as a key turning point, after which 'emotion' became 'a theoretical keyword at the heart of modern psychology' rather than only a synonym for passions or feelings.[13]

Once developed, James' ideas about the emotions – as the mind's perception of bodily response – became the influential James-Lange theory and were quick to cross the Atlantic.[14] These important developments in medical research were, however, rooted in psychology and had little impact on sexual forensics. Psychologists were absent from the courts and on the stand medical witnesses generally did not engage with the nature of relationships between body and mind. Pre-trial statements paid little attention to relationships between particular emotional signs and signified emotions, or to the processes by which emotions came to be embodied. Instead of looking at emotional indicators through the lens of new psychological research, medical and lay witnesses drew upon social ideas about *appropriate* emotional responses for people of different age, gender and social class.

Looking in depth at a single sign, crying, provides insights into the complex process of interpreting emotions in court and the social presumptions that fed into this process. In theory working-class complainants of both genders could be subject to emotional outbursts, including crying, in the courtroom; stoicism and the 'stiff upper lip' were associated primarily with upper-class males, who rarely entered the police courts as complainants. In cases of sexual crime, however, concerns about age and gender outweighed the class dimensions of emotions. Lay and medical evidence about crying related first and foremost to children, then to adult females, both of whom were expected to be close to nature and emotional. The rarity of testimony about emotional states in males was in part due to the low number of men in court, who constituted just ten per cent of complainants. However, there was also no testimony on the *absence* of emotion in male complainants, which indicates that questions about their emotional states were less likely to be posed by counsel. Witnesses only described the tears of men in relation to prisoners, for whom crying had specific connotations as a sign of remorse. When a prisoner was accused of an assault on a Middlesex girl aged six, for example, the prisoner apparently cried when confessing to the crime. The complainant's father testified about the prisoner's alleged remorse:

> The doctor asked him if he had touched the child with his hand on his person – Def[endan]t said I have – The doctor asked him how long he had been in the habit of doing so. Def[endan]t said 'shortly after we came here' – he was asked if he had done it recently – he said he had. Def[endan]t said he was very sorry and cried.[15]

Although these tears came during the prisoner's interview with the doctor, they were not a medical concern. The medical witness in this case made no reference to crying, noting only that 'I asked him if he had used his Person to the child and he said he had – I asked him if he had recently – he said he had – he said he had been in the habit of doing it'.[16] There was also no comparable testimony for an adult male complainant.

In contrast, medical practitioners and lay witnesses regularly testified to witnessing children of both genders crying. This focus drew upon a belief that children would demonstrate basic emotional signs, such as tears, because of their generally limited development and lack of understanding about the implications of a sexual offence; before puberty, the emotions were thought to be basic and were not yet gendered. Witnesses described 12 young males as crying, or as having cried. Although this figure is a small proportion of the total number of witnesses, it demonstrates the clear age dimensions of testimony about male emotions. Tears did not represent a failure of masculinity in the young, as pre-pubescent boys were not expected to have developed the 'will' to control themselves. When a boy aged 12 was apparently assaulted in the cloak room of Victoria Station in 1877, he stated that 'I began to cry when he let go and the policeman heard me ... I have never been punished for telling falsehoods' and a policemen testified that 'he was crying ... he said "he's been feeling me about and telling me how they make babies" '.[17] Alongside testimony about the complainant's innocence and ignorance, these tears represented the simple instinct of a child. Crying also constituted the majority of witness evidence about (possible) emotional signs for girls under the age of 13.[18]

The expectation that children were naturally emotional meant that the courts placed an onus upon them to demonstrate such emotions in the aftermath of a crime. They focused not only on the presence of tears, but also on their absence. In 1880, for example, a nine-year-old girl testified that she went to the beach at Weston-Super-Mare where a man 'pulled up my clothes' before giving her a half penny 'after he laid on me'.[19] A passer-by witnessed 'the little girl laying down and the prisoner laying on top of her'.[20] When the prisoner's solicitor asked some further questions, the eyewitness testified that '[t]he little girl in the presence of the prisoner said he was hurting her. She was not crying or resisting at all when I saw her'.[21] This testimony implicitly linked the girl's lack of tears to her lack of resistance. In the evening, the girl's mother took her to see the Resident House Surgeon of the Weston Super Mare Hospital. The surgeon testified, again under cross-examination, that '[t]he child was not excited. Perfectly collected'.[22] Proving the absence of emotion was a viable defence strategy in such cases, due to the widespread social and medical belief that girls should demonstrate primitive emotional responses to a sexual crime even in the absence of understanding; children were not expected to have developed the self-control to hide their emotions by the age of nine. Medical testimony that a girl was 'perfectly collected' in the aftermath of a sexual act raised doubts about her reliability or even, in the presence of an eyewitness and evidence that she took money from the prisoner, her complicity.

Drawing attention to the absence of emotion was commonly a defence strategy. A Middlesex trial from 1881, involving a complainant aged 11, further indicates the significance of such testimony. In pre-trial records, which either the magistrate or trial participants annotated, there is a pencilled mark underlining the complainant's statement that 'I did not call out. I was not crying' after an alleged assault.[23] Again, the prisoner's defence elicited this statement and implicitly aligned the girl's failure to cry with her lack of resistance. Children who maintained their normal demeanour in the aftermath of a crime did not adequately perform their roles as delicate, emotional victims. Medical practitioners reinforced such age-based stereotypes, as they focused on tears in relation to young children. Medical testimony carried some weight in court, although lay witnesses spoke on the presence or absence of tears with even greater frequency and played an equally important role in fuelling the 'emotional child' as a trope of victimhood.

Crying was evidently an important sign, particularly for the young, but like many emotional indicators had no straightforward meaning. Despite the perceived significance of the emotions as signs of mental disruption, lay and medical witnesses alike found it difficult to label and interpret them. There was no single or simple correlation between any emotional sign and emotional state. Tears were not only associated with emotion by the nineteenth century, but could also be social performances linked to moral or religious sensibilities.[24] Crying was also widely viewed as a possible response to pain, making it difficult to differentiate between the two in cases of penetrative or violent sexual crime. There was actually little onus upon witnesses to expand upon the meaning of crying as a sign. For the law, crying was corroborative evidence of an assault irrespective of whether it marked emotions or pain. Furthermore, expert witnesses could legally only give opinions 'within their own science'. As the emotions did not fall clearly within the remit of 'science', with medical practitioners and moral philosophers alike making interventions on the subject, medical witnesses only interpreted the meaning of tears when a purely physical matter such as pain was the cause. In Middlesex in 1878, for example, a medical witness testified that '[t]he child was crying from pain. Its underclothing was stained with blood ... [The injury] would cause pain and make it difficult for a child to walk'.[25] The complainant was only five years old and there is no evidence that she informed the medical practitioner that she was crying 'from pain'. The medical witness used his knowledge of her injury to give a professional opinion 'within his own science' on the reason for her behaviour. In this case the girl's tears were physically grounded and medically relevant.

There was a clear somatic (physiological) base to emotions in medical theory, but in court medical witnesses were often reduced to the status of 'ordinary' witnesses who could observe but not explain signs. Neither medical nor lay witnesses gave clear interpretations of potential emotional indicators such as crying. Lay witnesses made statements such as 'she was crying very much and appeared frightened' or was 'crying and in a most excited state', while medical witnesses made comments such as 'she was crying and in distress and nervous but not particularly agitated'.[26] The use of 'and' instead of 'because' meant that links between bodily signs and emotional states were ambiguous. The vague nature of such testimony potentially negates the historical value of crying as a sign because, in the words of Thomas Dixon, 'a tear on its own means nothing'.[27] However, medical and lay witnesses drew implicit links between tears and emotional states by listing them together in testimony. Witnesses listed tears alongside emotional states including agitation, downheartedness, nervousness, alarm, worry, mental disturbance, distress, fear and excitement. Crying was thus a vague but not meaningless sign. It connected to a number of subtly different emotional states, including high and low emotions, as well as to pain. Although crying represented no single emotion, for the courts at least, it operated as an umbrella indicator of short-term change in behaviour and feelings after an alleged offence.

The ambiguity of possible emotional signs was not limited to tears. Figure 5.1 demonstrates the range and inconsistency of connections between bodily signs and emotions, depicting the most common physical signs that witnesses linked to emotional states implicitly using

Emotions: Medicine and the Mind 139

	Red-faced	Quiet	Pale	Crying	Faint	Flustered	Shaking
Agitated	0	2	0	10	1	0	1
Low/Downhearted	0	2	1	1	0	0	0
Nervous	0	0	0	1	0	0	1
Alarmed	0	0	1	2	0	0	0
Worried	0	0	1	0	0	1	0
Disturbed	0	0	0	1	0	0	0
Upset/Distressed	1	2	1	32	3	0	2
Frightened	2	11	8	48	3	2	5
Excited	2	3	4	33	2	1	4
Explicitly not emotion–exhaustion	0	9	0	0	6	0	0
Explicitly not emotion–pain	1	0	1	16	0	0	0

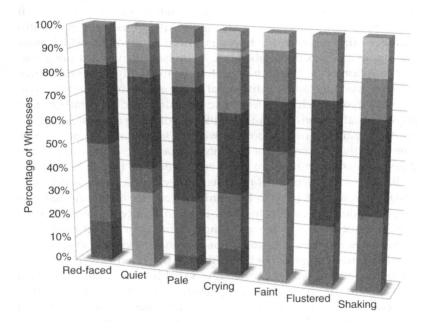

Figure 5.1 Links between bodily signs and emotional states
Note: Implicit links between bodily signs and emotional states on the basis of conjunction use. Statistics represent all witnesses, therefore some cases are represented twice

conjunctions such as 'and' and, less commonly, 'because'.[28] The limited number of such cases prohibits a systematic comparison of witness type, age, gender, time and place, but there may have been some patterns along these lines. In relation to age, for example, children represented a 'primitive' evolutionary stage similar to that of animals; this attitude was particularly prevalent among those who accepted Darwin's work.[29] The emotions of children were thought to be instinctive and basic, such as anger or grief, and required no understanding of any crime that had been committed. Based on such physiological theories, medical witnesses were more likely to link children's physical signs to fear than to less 'primitive' emotional states such as downheartedness. They also only described children below the age of 14 as 'frightened', dropping this term in favour of more complex 'nervous' states for older complainants.

Even if such trends are taken into consideration, it is clear that bodily signs were never mapped directly onto particular emotional states. A lack of general clarity resulted from the fact that every sign was connected to at least three emotional states and in some instances, such as fainting, to physical as well as emotional factors. This graph also excludes other signs and behaviours linked to emotion, which were even more complex and ambiguous. Witnesses often linked the physical disruption of a woman's hair or clothing implicitly to emotional distress, for example. Some behavioural responses such as 'falling down' in relief upon reaching safety were also dramatic acts, rather than physiological responses to a threat like shaking or paleness. Such bodily states and performed emotions were an important part of the 'real rape' narrative, but could not be easily mapped onto a single emotional state or theorised in terms of physiology. However, this ambiguity was no hindrance to the value of these signs as evidence; bodily and behavioural signs did not have to be clearly or consistently connected to a single emotion to be useful in court. The significance of the signs lay in their correspondence to age and gender, more than to particular emotional states: children and women needed to demonstrate some degree of behavioural disruption, in line with their perceived sensitive natures.

Lay and medical witnesses alike gave variable interpretations of signs such as crying, fainting and shaking. Medical practitioners showed little awareness of new research on the body-mind relationship and instead drew upon relatively vague assumptions about the emotional natures of complainants at particular ages. Overall, medical testimony on emotional signs helped to construct and reinforce age-based stereotypes of emotional victimhood, but medical practitioners acted as 'ordinary' rather than 'expert' witnesses in doing so. Some emotional *states*, however, actually constituted medical diagnoses. These emotions provided a rare opportunity for medical practitioners to give opinions in court as 'men of science'. The most significant of these were 'nervous' states, which were again grounded in stereotypes of age- and gender-based emotionality. Despite the continued influence of middle-class models of ideal femininity, medical testimony on the nerves as a somatic concept did change and respond to new medical research in a way that testimony on emotional signs and the body-mind relationship did not.

Nervous states had particular medico-legal and medico-moral significance. They were not only 'scientific' in nature, but could also be observed beyond the immediate aftermath of an alleged sexual offence. They also had a general significance in the nineteenth century as part of the distinction between 'emotions' and earlier ideas about 'passions'.[30] As Keith Oatley notes, ' "emotion" is a term from literary and scientific clusters that became prevalent only during the nineteenth century. These clusters include other terms such as expression, nerves, viscera, and brain'.[31] As emotions and the nerves overlapped in such scientific 'clusters', nervous conditions permitted a small sphere of medico-legal expertise on the emotions; this expertise anticipated the later rise of psychiatry and psychology in the courts. Although Figure 5.1 indicates that nervous states were rarely diagnosed, nervous conditions were actually one of the few emotional states that medical witnesses could diagnose with such confidence that they made little reference to particular signs or symptoms.

Nervous conditions were thought to affect women primarily after puberty, when gender differences emerged and when menstruation wreaked havoc with the body's energy. In 1885 the *British Medical Journal* noted that puberty was the most 'emotional and plastic period of life', marked by the emergence of unsettling sexual feelings in the absence of the requisite knowledge to control them.[32] It was at this age that women apparently became more emotionally volatile, in consequence of the disrupting influence of menstruation, and moved from the primitive emotions of childhood into more complex emotions linked to the nervous system. The connection between nervous energy and menstruation can be traced back to early conceptions of hysteria; the etymology of hysteria, in 'hyster', related directly to the womb. In 1900, the *British Medical Journal* implied that female emotionality was still linked to the sexual organs when it printed a gynaecologist's statement that women were 'the weaker or more emotional sex'.[33] Early modern theories of the 'wandering womb', witchcraft and early models of 'hysteria' thus clearly laid the grounds for Victorian ideas about female 'weakness' and sensitivity, although were not direct antecedents of the 'nerves'.[34] As the female nerves were apparently easily over-stimulated and quick to suffer from depleted energy, medical writers repeatedly emphasised that female nervous systems were sensitive. W. S. Playfair, Professor of Obstetric Medicine, writing in 1897, noted that:

> What I think we ought especially to bear in mind is the highly sensitive nervous organisation of the human female, the development of the emotional element which distinguishes the woman from the man ... Up to the time of puberty there is comparatively little difference between the sexes which distinguishes one sex from the other. As soon, however, as the great function of menstruation is established, the sexual period of a woman's life begins, which must henceforth dominate her whole existence ... Judiciously managed, she may be so trained that she will be able to meet successfully the strain on the nervous system which she has so often to cope with in after-life, such as the duties of a wife and mother, or the struggle with domestic anxieties, worries, and sorrows which are rarely altogether absent.[35]

The impact of menstruation on the female body was apparently comparable to the state of animals on heat and made a woman vulnerable to her natural, emotional state. As Helen King notes, in the late-nineteenth century 'the nervous system was considered to be under great stress, making a woman irritable, and prone to the passions of jealousy, anger and vengeance'.[36]

Despite the generally emotional nature of females, there was a medical and social expectation that maturity brought with it the development of female 'modesty'.[37] In consequence, respectable women would apparently control the outward indicators of their emotions after puberty. A rape or indecent assault, however, would make it impossible for any woman to hide her 'natural' emotional state. General theories of the emotions therefore fed into a particular construction of female victimhood. The 'real' post-pubescent female victim would be modest and emotionally measured in everyday life, but demonstrate an emotional response to an assault – most likely in the form of a nervous state. As females grew older they were increasingly expected to demonstrate a response to an assault without being *over*-emotional. This framework acknowledged the mental dimensions of sexual crime but also set up an expectation that a 'normal' woman would and should be unsettled for hours, days or even weeks after an assault.

The expectation that women would show emotional responses to crime was evident in medical testimony, which made reference to 'nervous' states in the aftermath of an alleged offence throughout the late-nineteenth century. The widespread belief that nervous states were fundamentally bodily in nature meant that testimony on these subjects fell within the remit of a general practitioner, rather than requiring a psychiatrist to testify in court.[38] Although lay witnesses testified more regularly on the emotions in general, medical witnesses dominated testimony about nervous conditions and shock.[39] They played an important role in medicalising these emotional states within courtroom scripts. In line with broader medical theories on the nerves, such testimony was limited to female complainants and medical witnesses paid more attention to nervous conditions as females grew older.[40] In the period 1850–1914, the first reference in court to a woman's 'nervous' condition was in 1855 in Somerset. The complainant, a married woman of unspecified age, stated that while walking home she met:

> [A] man who is the prisoner ... I replied good night and being frightened tried to pass him. He was standing before me: he caught me by one hand first and then took the other hand and held both behind me with one hand and with the other hand he pulled up my clothes before I struggled and kicked him in the legs. I screamed out: he then pushed me against the bank ... He then put his fingers into my private parts ... I think I was struggling with him for ten minutes during this time.[41]

Upon arriving home, her husband testified, she was 'in hysterics during the night and I could not get any understanding from her so as to put sense in her words: in the morning she told me that the man had thrown her down and made work with her private parts'.[42] These witnesses both testified on the subject of emotions, as the complainant described her fear during the assault and her husband testified about her 'hysterics' afterwards. Both of these accounts corroborated the complainant's story and provided a model of 'real rape', involving an assault by a stranger, a struggle and an emotional victim; the prisoner was sentenced to 12 months hard labour. Most evidence of this kind came from lay witnesses, who were the most likely to see a victim in the immediate aftermath of an assault. However, medical witnesses also had a particular role to play in medicalising emotional states. In the above case William Murray Silke, a local surgeon, stated that the complainant 'was in a frightened, nervous state' when examined after the alleged offence.[43] Such testimony indicated the nervous roots of emotions, whereas the lay witnesses described only the manifestations of these emotions in the form of 'hysterics'.

The 'nervous state' represented a general unsettling of the body and mind. As British philosopher John Stuart Mill stated in the 1862 edition of *A System of Logic*, mental states were likely caused by nervous states, but 'we are wholly ignorant of the characteristics of these nervous states; we know not, and at present have no means of knowing, in what respect one of them differs from another'.[44] The language of general 'nervous states' or 'nervous conditions' continued over subsequent decades, but responded to new research of the 1860s. By this decade, nervous conditions had become a way to articulate concerns about the demands of modern life, beginning with technology and extending to urban immorality and sexuality. In the 1860s, with the publication of surgeon John Eric Erichsen's book *On Railway and Other Injuries of the Nervous System*, 'railway spine' emerged as an important diagnostic category explaining the effects of train accidents on nervous states.[45] 'Railway spine' was a long-term nervous condition resulting from physical trauma, with both physical and emotional effects. This diagnosis marked a new recognition of the potentially long-term emotional effects of a traumatic event, and a growing emphasis on the body-mind relationship. Although 'railway spine' was primarily organic in nature, scholars have long thought of the diagnostic category as a step towards the concept of psychological 'trauma'.[46] This growing medical interest in the link between traumatic events, emotions and the nerves would come to influence medical responses to sexual crime. 'Railway spine' also connected to a wider contemporary interest in nervous shock, such as Edwin Morris's *A Practical Treatise on Shock* of 1867, which could result both from physical trauma and from strong emotions.[47] Some medical writers highlighted specific mental concerns as factors contributing to shock, including a sense of helplessness, while others focused on links between nervous shock and other nervous conditions such as hysteria.[48]

In the 1860s George Beard also published his internationally influential theories on the modern lifestyle disease of 'neurasthenia'. Neurasthenia, Beard argued, resulted 'from any causes that exhaust the nervous system' including childbirth, family responsibilities, business, sexual excesses and the abuse of narcotics.[49] Although for Beard neurasthenia was a problem of 'civilized' people, his work drew upon broad medical studies of the nervous system and nervous energy. Victorian concerns about nervous energy were based on the belief that the demands of modern life would deplete the body's finite reserves of energy, or what Beard described as 'nervous force'.[50] This 'nervous force' apparently existed in, and was lost from, the blood and sperm but could also be depleted through high emotions, shock or physical trauma. These medical theories about the links between emotions and nervous states had a particular social significance and should not be reduced only to the status of 'early' ideas about trauma. The links between energy and blood or sperm explain the particular concerns around 'wasting' sperm through masturbation and the influence of menstruation on female nerves. Medical literature on railway spine and neurasthenia did not focus on women, but connected broadly to gendered concerns about the links between nerves and the emotions. By the turn of the century nervous 'shock' took interconnected forms, which involved the loss of nervous energy or excitability.[51]

In line with this new medical research, responding more quickly than to other developments in psychology and physiology, medical witnesses began additionally to use the more specific language of 'depressed' nerves and 'nervous shock' in the 1870s. Such testimony continued to focus on females close to the expected age of puberty or above it. When 13-year-old Sarah Palett was minding a baby for her aunt, in Middlesex in 1879, her uncle allegedly 'held me with one hand & put down his trousers with his other. He got as far as my legs but could get no further'.[52] She was examined by a local surgeon, who apparently found her 'in a nervous & frightened state. She was suffering from a slight nervous shock'.[53] In the same year, another surgeon found a woman aged 33 to be suffering from 'great nervous depression' in the aftermath of an alleged sexual assault.[54] Although nervous states had been discussed in court throughout the nineteenth century, the introduction of 'nervous *shock*' as a concept in the 1870s had particular and novel implications. It no longer referred to a 'nervous' state in the immediate aftermath of an alleged assault, but to a longer-term form of emotional disruption grounded in the body. This change marked a shift towards acknowledging the longer-term, albeit not yet chronic, impact of sexual crime upon a female's mind rather than only her morals. Further explanatory frameworks for the emotions emerged in the early years of the twentieth century, relating to the psychological (G. Stanley Hall's influential dissemination of the notion of adolescent 'storm and stress' in 1904) and hormonal ('discovered' in 1905) origins of emotions at puberty.[55] However, these theories were nascent and supplemented rather than replaced nervous models of emotions.

The language of shock gradually entered general society. In 1883 a married complainant from Middlesex testified that 'I went to the Doctor the same day I came here. I am pregnant and the shock to the system caused by this assault induced me to go'.[56] In 1898 *Trewman's Exeter Flying Post* reported that another complainant 'was in bed for two days owing to the shock on her nerves from the struggle with the prisoner'.[57] These lay references to shock were outnumbered by medical discussions of the same, but indicate that new medical models of nervous 'shock' entered public consciousness. Lay witnesses also paid great attention to states of so-called 'excitement', which overlapped conceptually with medical ideas about the excitability associated with nervous shock. In the nineteenth century the extremely common lay rhetoric of 'excitement', according to the *Oxford English Dictionary*, meant 'stirred by strong emotion'.[58] 'Excitement' was particularly widespread in lay testimony because it had meaning beyond the purely emotional. Witnesses used the term in association with physical exhaustion and many prisoners put forward the defence that they were 'excited' by drink, to explain their actions. Despite these broad connotations, there was a clear emotional dimension to the term. In 1885 when a woman was allegedly assaulted in a Middlesex cab, a policeman testified explicitly that 'she was excited from being pulled about more than from drink'; 'excitement' from alcohol consumption was potentially negative evidence about a woman's moral character, but purely emotional 'excitement' corroborated the charge.[59] In 1898 in Somerset, the mother of one complainant linked 'excitement' to emotional states when she stated that her 16-year-old daughter was 'a little excited or frightened'.[60] Medical testimony about 'nervous depression' also shared characteristics with lay testimony about complainants being 'downhearted' or low-spirited. When a Devon doctor referred to depression as making a female complainant 'nervous and dull' in 1870, his language was similar to that of a Middlesex case in which a young girl's mother testified in 1876 that 'for the last 10 days the child has been very different, so dull'.[61] Although medical language had a particular power in court, it overlapped conceptually and occasionally discursively with lay testimony on the emotions.

Due to the limited sample and great range of factors in any one trial, it is difficult to determine the extent to which evidence about emotions influenced outcomes. The fact that lay and medical witnesses alike testified on the emotions indicates that they expected the subject to have some weight in court. However, evidence that a complainant was 'nervous' or 'excited' seemingly only corroborated a charge if she was not *generally* emotional in nature; over-emotionality could apparently lead to nervous states such as hysteria, which in turn made a complainant unreliable and negated the value of emotions as evidence. Females had to tread a very fine line between showing insufficient emotion and too much emotion in order to be seen as a reliable witness. As one physician wrote in the *British Medical Journal*, in 1866, 'those who are readily accessible to impressions coming from without, who feel acutely and are liable to strong emotions, are certain to become hysterical if made to suffer mental agony or prolonged physical pain'.[62] Such comments supported hysteria as a sign or symptom of assault to a degree, by relating it to 'mental agony or prolonged physical pain'. However, over-emotional women were also deemed untrustworthy. Medical literature and newspapers highlighted a tendency for over-emotional and hysterical females to lie, particularly in relation to alleged offences against medical practitioners themselves.[63] Witnesses in court clearly tried to remove themselves and their relatives from potential allegations of over-emotionality. In Devon in 1899, when a 13-year-old servant girl alleged that she had been subjected to a sexual assault, her father testified that 'she is a little delicate but not in the habit of fainting and not of excitable temperament'.[64]

Defence teams were quick to utilise the perceived link between hysteria and false allegations. Even when medical practitioners denied such links, as they did occasionally at trial, they were questioned closely on the implications for a complainant's reliability of suffering 'hysterical attacks'.[65] Defence counsel also linked hysteria with lies when they summed up a case for the jury. When London's *Morning Chronicle* reported on a Middlesex case of alleged indecent assault on a 15-year-old girl, who 'looked much older', in September 1856 it stated that:

> Mr Ribton, in his address to the jury, urged that there might be a great deal of falsehood mixed up with a little truth in this matter, to give the story the appearance of probability; that the girl might have gone voluntarily to the prisoner's house after having sent her sister out; and that she might have been suddenly seized with a hysterical fit, though she never had one before. She might, he submitted, have inverted the greater part of the story to screen herself from the blame of being there.[66]

The hysterical fit had been the primary symptom of hysteria in the early-nineteenth century and represented a physical response to sudden and strong emotional stimuli.[67] This focus on the hysterical fit, still evident in 1856, meant that hysteria was understood in terms of particular symptoms rather than yet in terms of the 'hysterical woman', prone to lying. However, the 'hysterical fit' was already linked implicitly with 'falsehood' by the mid-nineteenth century and laid the groundwork for growing links between hysteria and false charges. Medical theory on hysteria moved away from a focus on the 'hysterical fit' towards the hysterical personality type in the late-nineteenth century. This shift consolidated and strengthened the tendency to distrust over-emotional or nervous women. Medical writings thus began increasingly to consider a 'hysterical personality' marked by egocentricism and the lack of emotional control, a notion that would heavily influence Freudian thought.[68]

Medical literature paid increasing attention to the subject of hysterical women over the course of the late-nineteenth century. This shift was not only the result of changing physiological and psychological theory, but also became a way to articulate medical anxieties about false charges against the profession in the wake of the Criminal Law Amendment Act. In 1885 the *British Medical Journal* reported that a 'Dr. Trestrail, of Aldershot, was charged with having committed rape on an hysterical patient ... the jury summarily terminated the proceedings by returning a verdict of "not guilty" '.[69] In the aftermath of a similar case, tried in 1894, the *British Medical Journal* reprinted the comments of a London newspaper that '[d]octors are peculiarly exposed to these accusations, and their only protection is that somewhat uncertain quantity, the common-sense of a jury. It might, perhaps, have a salutary effect if the Public Prosecutor were occasionally to prosecute one of these hysterical ladies for perjury'.[70] Medical testimony on the nerves and hysteria became less common after the turn of the century, however. In London and the South West, the final medical and lay references to 'nervous shock' came in 1898. The final reference to 'nervous' conditions more generally was in 1903. The absence of any testimony about 'nervous' states between 1903–1914 may indicate the declining interest in such ideas before shell shock in World War One.

Medical thinkers questioned and redefined the body-mind relationship in the late-nineteenth century, but the specific physiology of emotions was irrelevant in court. Judges, juries and solicitors wanted only to know whether a complainant had acted in an unusual way in the aftermath of an alleged offence, both in relation to their own particular temperament and to the emotional norms of women and children. Evidence about the presence or absence of signs placed an additional burden upon women and children, who needed to prove their 'victimhood' not only with physical harm, but also with a mental shift in the aftermath of a crime. This emotional response also needed to become more complex with age to remain convincing to the court. After puberty women required witness testimony and a courtroom demeanour that together demonstrated the right degree and type of emotion, without any disruptive medical diagnosis of 'hysteria' or over-emotionality. The female 'victim' was not only created by a crime, but was a type of personality marked by emotional vulnerability and modesty in everyday life.

Case Analysis: Emotions, 1889

The Case: This case builds upon the chapter's analysis of female over-emotionality and its judicial implications. It shows, in greater depth, how the value of emotions diminished when a complainant had a generally emotional nature or, as in this case, was diagnosed as hysterical. The case also demonstrates the role of medicine in constructing and fuelling concerns about the hysterical female in the late-nineteenth century, particularly when the medical practitioner was the accused party. Its testimony has general relevance for the question of emotions as evidence in court, particularly the notion that emotions needed to mark a change from emotional equilibrium (or the appearance thereof).

The Prisoner: William Jones, age unrecorded, medical practitioner.
The Complainant: Annie Louisa Parsonson, aged 18.
The Complaint: Indecent assault (consent a defence).
The Pre-Trial Statement: Depositions taken 11 April 1889, Edmonton Police Court.[71]

Annie Louisa Parsonson: I am 18 years old and unmarried and am living as a domestic servant with Mrs Woolford at 194 Fore St, Edmonton … I have lived there 4 months. I was at a fete at Symmes Park last year with some friends who knew def[endan]t and I met him there on that occasion. On 1st April 1889 about 6.30 pm I called on the defendant at his residence in order to consult him professionally. I was shown into a consulting room. Defendant came into the room and closed the shutters of the window, which looks across a garden in the street. I told him I had come because I had a cold on my chest. He sounded my chest. I was standing up having risen when he came into the room. He said I had a very bad cold and would make me up some medicine. He then left the room. He shortly afterwards returned to the room and gave me a bottle of medicine. I then said that I had pains in my head, which I thought were caused by 'irregularity'. He asked me to shew [sic] him my breasts. I undid my dress and shewed [sic] him my breasts. He said 'oh you're in the family way'. I said 'I am not'. He then said 'let me examine you' … Then he examined me by forcing his finger into my private parts. He then said 'some cock has been here, there's room for my finger and there's room for anything'. He then kissed me and said 'your maidenhead is broken'. He used his right hand to examine me. His left hand was round my waist at the time. I got up and said I don't know what you mean my mother has never told me about such things … My mother was not in when I got home. I went up into her bedroom and threw myself on the bed in hysterics. My little sister went and fetched my mother and directly she came I told her what had occurred. She took me at once, as soon as I had left off crying, to Dr Scott's at railway approach. My mother told him I had seen Dr Jones and Dr Scott examined my chest and gave me some medicine. Dr Scott had a private interview with my mother while I went into another room. I know I am not pregnant. I have never been in a position which would lead to pregnancy ... I said to him 'I have seen nothing for two or three months'. That is the way I referred to my being irregular and I said to him that perhaps he might give me something in my medicine to bring me on ... I did not ask him to leave me alone, because he was a doctor and I thought a doctor was allowed to do what he likes. It was after I told my mother that I found a doctor was not allowed to do what he likes ... I went to Dr Jones because I thought as he attended my friends he would take an interest in me. I never had fits. I am still in the same situation.

Jane Elizabeth Parsonson: I am the wife of Robert John Parsonson, carver and gilder, and lived at 48 Beaconsfield Road, Edmonton. Prosecutrix is my second daughter. She is in service ... I was expecting her at home in the evening of 1st April because it was her evening out. About 7.15 or 7.20 pm on that evening I was at my next door neighbours when I was called home by my little daughter. I went into my bedroom and found prosecutrix lying across the bed crying, I may say screaming. I coaxed her and asked what was the matter. She made a complaint about the def[endan]t and on hearing it I immediately took her to Dr Scott and told him what prosecutrix told me. He examined her bosom and chest. I have had 12 children. I am aware prosecutrix has been irregular as regards health. I have known her to go as long as 6 months. I feel satisfied there is no ground for believing her pregnant ...

Walter Frederick Scott: I am a Licentiate of the Society of Apothecaries and practise at Lower Edmonton. On 1st April about 8 pm prosecutrix came to me accompanied by her mother. I examined the girl's chest with reference to the condition of her lungs and I prescribed for a cold. After that prosecutrix's mother made a further statement to me, upon which I made a further examination of prosecutrix's bosoms. I came to the conclusion that the girl was not pregnant. From having attended the girl before and from the statement made to me I arrived at the conclusion without having occasion to make a digital examination. I have heard the evidence given by prosecutrix. The method of examination alleged to have been employed is not the usual method. (cross-examined) I certainly should not have a girl on a chair for the purpose and I should object to make such an examination except in the presence of another female. The girl has since become unwell in the ordinary way. Apart from that and from what I saw on 1st April

I formed an opinion she was not then pregnant but could not then have sworn as a fact that she was not. (by Bench) I am satisfied that prosecutrix has been unwell since 1st April in the ordinary way and not in the way of an early miscarriage.

For the defence

William Jones (the Defendant): I am a Surgeon in practice at Edmonton. I have been qualified ten years, during the last 9 years of which I have practised at Edmonton. About 7 pm on 1st April I was at home upstairs when a message was brought to me by the page boy. I went downstairs and found prosecutrix in the consulting room. I said 'good evening, what can I do for you?' She told me she had a bad cold. I then examined her breast outside her dress. She then told me that she perspired at night. I then told her to undo her dress so that I could examine her chest carefully. As she was doing so she told me she had a pain in the left side and I told her to undo the top of her stays to enable me to examine the apex of her heart. She unfastened the top of her stays and I examined her heart. I told her the action was quick but there was no heart disease. She then told me she had not seen anything for 2 or 3 months and said can you give me something to bring me on. I looked at her breasts and saw blue veins there. I then put the following questions to her: 'are you engaged?' To which she replied that she did not know what I meant. I asked: 'has any young man taken liberties with you?' She again said she did not know what I meant. I then said to her 'I'm afraid you've gone wrong' ... I then left the consulting room and went to the dispensary and a bottle of medicine was made up by my wife under my direction in the dispensary ... She went immediately after I gave her the medicine ... In my opinion the fact of a girl being irregular would be likely to bring on hysterical fits. Hysterical people will make any charges against anybody. Their imagination is great. I have only made one delicate private examination since I have been in my present house. On that occasion I took the patient upstairs and made the examination in drawing room in my wife's presence. I never heard the slightest suggestion of impropriety made against me as a medical man or otherwise. I see patients from 7 till 9 in the evening. I had no recollection of having seen prosecutrix before.

Emily Elizabeth Jones: I am the wife of the def[endan]t ... The girl was in the house less than 10 minutes. (by bench) My husband did not close the consulting room when he came for the medicine.

Alfred Foster: I am a page boy at Dr Jones' at Edmonton. I remember letting prosecutrix in on 1st April . It was about 7 pm ... I heard him come down about two minutes afterwards ... (cross-examined) I did not go back to consulting room or tell you after taking message to def[endan]t.

The Verdict: No Bill.

Significance for Sexual Forensics: The idea of a 'hysterical personality' provided a useful defence in this case from 1889, in which a surgeon used hysteria as a defence when accused of assaulting an 18-year-old female patient. Although a defendant rather than a forensic witness, he took advantage of the widespread mistrust of hysterical witnesses when testifying that '[h]ysterical people will make any charges against anybody. Their imagination is great'. Such an approach differed from that of lay witnesses, for whom 'hysterics' or being 'hysterical' were behaviours or bodily responses to an assault not diagnoses. This practitioner referred not only to hysterical 'fits', but also to a hysterical *person*. He therefore spoke of the woman's general character rather than her short-term mental state. His description of the girl as a 'hysterical' person had very different implications to her own testimony, in which she described being in 'hysterics' after the alleged assault. The girl drew upon the lay language of emotions, whereas the defendant drew upon more medicalised language that pathologised the complainant's behaviour and personality.

The prisoner linked the 'hysterical personality' to a number of physiological symptoms. Most significantly, he described her general hysteria as a symptom of menstrual irregularity. Despite acknowledging the variability of menarche, the main medical journals emphasised that absent or irregular menstruation was a possible cause for concern at the age of 18.[72] As menstruation was so closely connected to energy circulation and female nervous states, suppressed and irregular menstruation were thought to be possible causes of hysterical fits. The prisoner may have been euphemistically referring to pregnancy when he spoke of 'girls being irregular', or to the wide medical literature on links between 'suppressed' and 'irregular' menstruation and heightened emotions or insanity.[73] Both shed doubt on her character, particularly when combined with the prisoner's claims about the girl's unchastity.

The medical witness for this case, in contrast to the medical prisoner, found no pregnancy and made no reference to the complainant's hysterical nature. However, his testimony that the girl was unwell in the 'ordinary' way confirmed witness testimony about her menstrual irregularity. Although the use of 'ordinary' played down the significance of this irregularity, it had broad social and medical links with pathology and hysteria. Menstrual irregularity apparently occurred for a short time at times of 'crisis' such as puberty and menopause, but when such irregularity continued for older girls it represented a longer-term 'crisis' due to the erratic way in which energy was being channelled towards the womb.[74] The medical witness in this case, who was a long-term doctor of the complainant, made no attempt explicitly to support the prisoner's defence, but did so implicitly. The complainant and her mother also admitted to her irregular menstruation and, although they denied that she suffered 'fits', this physiological fact in itself ensured that the girl was seen as 'abnormal' and potentially nervous in nature.

The prisoner's line of defence tapped into a wider connection between hysteria, false charges and medical practitioners. It was successful, in part, because it connected to these widespread stereotypes. A report from *The Times*, which outlined the grand jury's dismissal of the case, indicates that judges and juries were easily swayed by the prisoner's 'false charge' defence. It noted that '[t]he chairman, in addressing the grand jury … particularly called attention to a charge against a Mr. William Jones, a medical man, who had been committed from Edmonton on a charge of indecently assaulting Annie Louisa Parsonson, and remarked that it had been said that these charges were very easily made, difficult to prove, and still more difficult to disprove'.[75] This statement echoed the words of seventeenth-century judge Sir Matthew Hale, who famously wrote that 'rape is a most detestable crime … but it must be remembered, that it is an accusation easily to be made and hard to be proved, and harder to be defended by the party accused, though never so innocent'.[76] The statement was reformulated and reiterated to such an extent that Bruce Macfarlane has cited it as the 'birth of a myth' that consolidated long-held anxieties about false claims within and beyond Britain for hundreds of years.[77]

Such 'myths' framed both the nature of evidence that reached trial and the decisions of judges and juries at trial. In this case the chairman, like the medical prisoner, shed doubt on the complainant's reliability. *The Times* newspaper report also made two statements that the case was without 'foundation'. It did not make direct reference to the complainant's alleged hysteria, but supported implicitly the prisoner's claims about her lack of reliability. The report stated that the chairman 'felt constrained to tell the grand jury that if nothing more than the facts as they appeared on the depositions was brought forward, he should, when the accused was tried, certainly advise the petty jury that it would be very unsafe to convict'.[78] His reference to the pre-trial statements or 'depositions' transcribed here, with their emphasis on the complainant's hysteria, indicates that this evidence informed the chairman's strong recommendation. The grand jury directly followed his advice and the case never reached trial.

Notes

1. 'The Police Courts', *Daily News*, 23 August 1856, 4, p. 4.
2. 'The Police Courts', p. 4.
3. 'The Police Courts', p. 4.
4. An 'emotionalization of law' has occurred in the criminal justice system; Susanne Karstedt, 'Emotions and Criminal Justice', *Theoretical Criminology* 6 (2002), 299–317, p. 299. There has also been an 'emotional turn' in historiography in the last two decades. See, for example, Thomas Dixon, 'The Tears of Mr Justice Willes', *Journal of Victorian Culture* 17 (2012), 1–23.
5. Joanna Bourke, 'Sexual Violation and Trauma in Perspective', *Arbor-Ciencia Pensamiento Y Cultura* 743 (2010), 407–16, p. 407.
6. Peter N. Stearns and Carol Z. Stearns, 'Emotionology: Clarifying the History of Emotions and Emotional Standards', *The American Historical Review* 90 (1985), 813–36, p. 813.
7. Judith Butler, 'Performative Acts and Gender Constitution: An Essay in Phenomenology and Feminist Theory', *Theatre Journal* 40 (1988), 519–31, p. 519.
8. See Chapter 4 on consent and resistance.
9. Some dispute the common claim that Freud abandoned his seduction hypothesis completely. However, whether Freud fully retracted his seduction theory or not, ideas about the prevalence of child sexual abuse and recovered memories were certainly not widely accepted until the late-twentieth century.
10. William Fleming, *A Manual of Moral Philosophy* (London: John Murray, 1867), p. 29.
11. William James, 'What is an Emotion?', *Mind* 9 (1884), 188–205.
12. Fay Bound Alberti, *Matters of the Heart: History, Medicine and Emotion* (Oxford: Oxford University Press, 2010), p. 149.
13. Thomas Dixon, ' "Emotion": The History of a Keyword in Crisis', *Emotion Review* 4 (2012), 338–44, p. 338.
14. 'Association Intelligence', *British Medical Journal (BMJ)*, 21 July 1900, 155–71, p. 162.
15. London, London Metropolitan Archives (LMA), Pre-Trial Statements, Thomas Swigg tried at the Middlesex Sessions on 27 April 1860 for attempted carnal knowledge, MJ/SP/E/1860/009.
16. London, LMA, Pre-Trial Statements, Swigg, MJ/SP/E/1860/009.
17. London, LMA, Pre-Trial Statements, Samuel Rawson tried at the Middlesex Sessions on 19 June 1877 for indecent assault on a male, MJ/SP/E/1877/013.
18. Taking all witness testimony about possible emotional signs (including signs such as red-faced, fainting, crying, trembling and quietness), crying was the subject of 57 per cent of testimony about emotional signs for girls aged between one and six years old, 58 per for girls aged seven to 13, 44 per cent for girls aged 14–20 and 37 per cent for females over the age of 21.
19. Taunton, Somerset Heritage Centre (SHC), Pre-Trial Statements, William Fogary tried at the Somerset Quarter Sessions on 30 June 1880 for indecent assault, Q/SR/720.
20. Taunton, SHC, Pre-Trial Statements, Fogary, Q/SR/720.
21. Taunton, SHC, Pre-Trial Statements, Fogary, Q/SR/720.

22. Taunton, SHC, Pre-Trial Statements, Fogary, Q/SR/720.
23. London, LMA, Pre-Trial Statements, Peter Minnella tried at the Middlesex Sessions on 23 August 1881 for carnal knowledge, MJ/SP/E/1881/031.
24. Dixon, 'The Tears of Mr Justice Willes'.
25. London, LMA, Pre-Trial Statements, Dennis Ryan tried at the Middlesex Sessions on 13 August 1878 for attempted carnal knowledge, MJ/SP/E/1878/025.
26. Taunton, SHC, Pre-Trial Statements, William Sparks tried at the Somerset Quarter Sessions on 18 October 1877 for assault with intent, Q/SR/709; London, LMA, Pre-Trial Statements, Alfred Froude tried at the Middlesex Sessions on 14 July 1882 for indecent assault, MJ/SP/E/1882/028; London, LMA, Pre-Trial Statements, Charles Perry not tried (no bill) at the Middlesex Sessions in September 1877 for indecent assault, MJ/SP/E/1877/021.
27. Dixon, 'The Tears of Mr Justice Willes', p. 4.
28. Insensibility and hysteria are excluded for clarity because they could be both emotional signs and states, depending upon the context and the witness. Similar signs have been merged, such as 'quiet' with 'could not speak', 'red-faced' with 'flushed', and 'trembling' with 'shaking'.
29. 'Two Lectures on the Physiology of the Emotions', *BMJ*, 11 April 1908, 853–58, p. 856.
30. Alberti, *Matters of the Heart*, p. 155.
31. Keith Oatley, *Emotions: A Brief History* (Malden, MA: Blackwell, 2004), p. 135.
32. 'Sexual Ignorance', *BMJ*, 15 August 1885, 303–4, p. 304.
33. 'Sixty-Eighth Annual Meeting of the British Medical Association', *BMJ*, 22 September 1900, 789–844, p. 791.
34. Susan E. Cayleff, ' "Prisoners of their own Feebleness": Women, Nerves and Western Medicine — A Historical Overview', *Social Science & Medicine* 26 (1988), 1199–208, p. 1200.
35. William S. Playfair, 'Remarks on the Education and Training of Girls of the Easy Classes at and about the Period of Puberty', *BMJ*, 7 December 1895, 1408–10, pp. 1408–9.
36. Helen King, *The Disease of Virgins: Green Sickness, Chlorosis and the Problems of Puberty* (London; New York: Routledge, 2004), p. 92.
37. Lucy Bland, *Banishing the Beast: English Feminism and Sexual Morality 1885–1914* (London: Penguin, 1995), p. 56. For an example of advice literature about the 'retiring' nature that came with menstruation see George Black, *The Young Wife's Advice Book: A Guide for Mothers on Health and Self-Management* (London: Ward, Lock & Co., 1888), p. 8.
38. Although psychiatry gradually gained a foothold in some of the larger courts, such as the Old Bailey, not a single alienist (psychiatrist) testified in Middlesex, Gloucestershire, Somerset or Devon in the period 1850–1914.
39. Medical witnesses made only 34 references to emotional states and possible signs out of 842 such references.
40. 'Nervous' conditions constituted 20 per cent of medical testimony about the emotional signs and states of girls below the age of seven; 33 per cent for girls aged 7–13; and 23 per cent for females aged 14–20. This latter figure was likely higher than it appears, because 38 per cent of medical testimony on the emotions of complainants of unspecified age – but often married or in service – also related to 'nervous' conditions.

41. Taunton, SHC, Pre-Trial Statements, John Coughlin tried at the Somerset Quarter Sessions on 4 January 1855 for assault with intent, Q/SR/610.
42. Taunton, SHC, Pre-Trial Statements, Coughlin, Q/SR/610.
43. Taunton, SHC, Pre-Trial Statements, Coughlin, Q/SR/610.
44. John Stuart Mill, *A System of Logic, Ratiocinative and Inductive: Being a Connected View of the Principles of Evidence, and the Methods of Scientific Investigation*, 5th edn, Vol. 2 (London: Parker, Son, and Bourn, 1862), p. 430.
45. John Eric Erichsen, *On Railway and Other Injuries of the Nervous System* (London: Walton & Maberly, 1866).
46. Edgar Jones and Simon Wessely, 'A Paradigm Shift in the Conceptualization of Psychological Trauma in the 20th Century', *Journal of Anxiety Disorders* 21 (2007), 164–175, p. 166; Ralph Harrington, 'On the Tracks of Trauma: Railway Spine Reconsidered', *Social History of Medicine* 16 (2003), 209–23.
47. Tim Armstrong, 'Two Types of Shock in Modernity', *Critical Quarterly* 42 (2000), 60–73, p. 61.
48. See Frederick Heaton Millham, 'Historical Paper in Surgery: A Brief History of Shock', *Surgery* 148 (2010), 1026–37; Jill Matus, *Shock, Memory and the Unconscious in Victorian Fiction* (Cambridge, Cambridge University Press, 2009).
49. George Beard, 'Neurasthenia, or Nervous Exhaustion', *The Boston Medical and Surgical Journal* 80 (1869), 217–21, p. 218.
50. Beard, 'Neurasthenia', p. 218.
51. Armstrong, 'Two Types of Shock'.
52. London, LMA, Pre-Trial Statements, Jesse Cave tried at the Middlesex Sessions on 8 July 1879 for indecent assault, MJ/SP/E/1879/014.
53. London, LMA, Pre-Trial Statements, Cave, MJ/SP/E/1879/014.
54. London, LMA, Pre-Trial Statements, George Hornsby tried at the Middlesex Sessions on 22 May 1879 for indecent assault, MJ/SP/E/1879/007.
55. G. Stanley Hall, *Adolescence: Its Psychology and its Relations to Physiology, Anthropology, Sociology, Sex, Crime, Religion and Education*, vol. 1 (London; New York: Appleton, 1904); Lynn Eaton, 'College Looks Back to Discovery of Hormones', *BMJ*, 25 June 2005, 1466.
56. London, LMA, Pre-Trial Statements, George Pilotel tried at the Middlesex Sessions on 10 May 1883 for indecent assault, MJ/SP/E/1883/022.
57. 'Devon Sessions', *Trewman's Exeter Flying Post or Plymouth and Cornish Advertiser*, 2 July 1898, unpaginated.
58. 'Excited, adj.', OED Online <http://www.oed.com/view/Entry/65796?redirectedFrom=excited> (accessed 15 March 2015).
59. London, LMA, Pre-Trial Statements, Thomas Byrne and Edward Newman tried at the Middlesex Sessions on 31 July 1885 for indecent assault, MJ/SP/E/1885/039.
60. Taunton, SHC, Pre-Trial Statements, John Miller not tried (no bill) at the Somerset Quarter Sessions on 6 April 1898 for indecent assault, Q/SR/791.
61. Exeter, Devon Record Office (DRO), Pre-Trial Statements, George Lange and John Sansome tried at the Devon Quarter Sessions on 29 June 1870 for attempted rape, QS/B/1870/Midsummer; London, LMA, Pre-Trial Statements, Joseph Vickery tried at the Middlesex Sessions on 23 August 1876 for attempted rape, MJ/SP/E/1876/017.
62. Julius Althaus, 'A Lecture on the Pathology and Treatment of Hysteria', *BMJ*, 10 March 1866, 245–48.

63. See the chapter's case analysis for an example of a surgeon, accused of assault, who used 'hysteria' as a defence.
64. Exeter, DRO, Pre-Trial Statements, Charles Easton tried at the Devon Quarter Sessions on 18 October 1899 for indecent assault, QS/B/1899/Michaelmas.
65. In 1870, for example, a Devon practitioner testified that a complainant was 'subject to violent hysterical attacks ... but not so as to prevent her knowing what she is about ... I have always known her to be a person of good character'. Exeter, DRO, Pre-Trial Statements, George Lange and John Sansome tried at the Devon Quarter Sessions on 29 June 1870 for attempted rape, QS/B/1870/Midsummer.
66. 'Middlesex Sessions – Saturday', *The Morning Chronicle*, 16 June 1856, unpaginated.
67. Carroll Smith-Rosenberg, 'The Hysterical Woman: Sex Roles and Role Conflict in 19th-Century America', *Social Research* 39 (1972), 652–78, p. 662.
68. Smith-Rosenberg, 'The Hysterical Woman', p. 662.
69. 'False Charges against Medical Men', *BMJ*, 2 July 1887, 44, p. 44.
70. 'False Charges Against a Medical Practitioner', *BMJ*, 14 April 1894, 817–18, p. 818.
71. London, London Metropolitan Archives (LMA), Pre-Trial Statements, William Jones not tried (no bill) at the Middlesex Sessions in May 1889 for indecent assault, MJ/SP/E/1889/022. Due to the length of the original pre-trial statement, it has been edited down to the most relevant testimony. Some punctuation has been added for clarity, but the meaning remains unchanged.
72. For example, see J. Henry Bennet, 'On Healthy and Morbid Menstruation', *The Lancet*, 3 April 1852, 328–29, p. 328.
73. Elaine Showalter, 'Victorian Women and Insanity', *Victorian Studies* 23 (1980), 157–81, p. 170.
74. Julie-Marie Strange, 'Menstrual Fictions: Languages of Medicine and Menstruation, c. 1850–1930', *Women's History Review* 9 (2000), 607–28, p. 616.
75. 'County Of London Sessions, May 25', *The Times*, 27 May 1889, 8, p. 8.
76. This quote replaces 'f' with 's' in words such as 'detestable' and updates the spellings of 'remembred' and 'tho' for readability, but is taken from the original text; Matthew Hale, *The History of the Pleas of the Crown*, rev. edition, vol. 1 (London: 1778 [1736]), p. 635. This book has been cited by numerous historians and scholars, including the particularly visible cases of: Sandy Ramos, ' "A Most Detestable Crime": Gender Identities and Sexual Violence in the District of Montreal, 1803–1843', *Journal of the Canadian Historical Association* 12 (2001), 27–48; and Keith Burgess-Jackson (ed.), *A Most Detestable Crime: New Philosophical Essays On Rape* (New York: Oxford University Press, 1999).
77. Bruce A. MacFarlane, 'Historical Development of the Offence of Rape', originally published in Josiah Wood and Richard Peck (eds), *100 Years of the Criminal Code in Canada: Essays Commemorating the Centenary of the Canadian Criminal Code* (Ottawa: Canadian Bar Association, 1993), p. 50 [pagination from online document] <www.canadiancriminallaw.com/articles/articles pdf/Historical_Development_of_the_Offence_of_Rape.pdf> (accessed 2 May 2015).
78. 'County Of London Sessions, May 25', *The Times*, 27 May 1889, 8, p. 8.

6
Offenders: Lust and Labels

The 'paedophile' emerged as a named object of concern in the late-nineteenth century as a result of sexologists' interest in the categorisation of sexual deviance. However, the label was rarely used in mainstream medicine. As Stephen Angelides notes, 'of principle concern to sexologists were sexual deviations with respect to the aim or gender of object choice, not the age of object choice'.[1] Only in the late-twentieth century was there an 'explosion of social panic' around paedophilia.[2] The belief in a specific type of offender prone to abusing children existed in the late-nineteenth century, but it gained little traction until a century later. In the absence of any clear-cut notion of 'paedophilia', or even of 'child sexual abuse' as a single type of crime, Victorian and Edwardian medical practitioners had no coherent idea of what motivated perpetrators of sexual offences against the very young.[3]

In her work on early modern rape, Garthine Walker notes that historians often make sweeping claims about the shift from rapist as 'everyman' to 'monster' under the influence of late twentieth-century feminism.[4] She rightly notes that there was a wide range of constructions of the offender at any given time, including both of these characterisations of the criminal, which differed according to factors ranging from the age of victim and offender to the type of crime. Walker's conclusions are equally applicable to the nineteenth century, in which medical practitioners and social commentators provided a range of different explanations for sexual crimes depending on their nature. The age of a prisoner and the complainant were crucial to understanding sexual crimes and to their construction as 'natural' or 'unnatural'. Middle-class ideas about masculinity and 'natural' male sexuality provided explanatory frameworks for crime when the offender had committed a heterosexual act against somebody of similar age. In contrast, many explanations for offences against the young operated either to desexualise the crime or to emphasise that the 'unnatural' misdirection of 'natural' lust was an abnormal consequence of circumstances, the prisoner's mental incapacity or his alcohol consumption. Only the 'pederast' emerged as a pathologised subject of concern in the nineteenth century because this offender committed two types of 'unnatural' offence: an offence against the young and a same-sex act.

The sexual offender was nearly always a man, as only a male could commit rape under nineteenth-century law. Writers of forensic medicine texts noted that women could technically sexually assault males, but such crimes would be grouped with sodomy under English law and rarely came before the courts. Francis Ogston, for example, noted in his medical jurisprudence textbook that female rape of males was 'little known in this country as a crime' although such cases had been brought before continental courts.[5] In the Middlesex Sessions and the south-west Quarter Sessions two women were tried for indecent assault between 1850 and 1914, but only as accomplices to men committing assaults on girls. The sexual offender was also rarely young. Only 16 prisoners under the age of 14 were brought to trial, all for indecent assault or attempted carnal knowledge rather than rape.

There was a legal assumption that boys were unable to sustain erections before the age of 14, although medical writers provided evidence to the contrary. Most medical jurisprudence texts agreed that 'in a great many cases [a boy under the age of 14] would be able to commit the crime'.[6] Despite such comments, which were widespread, the law on male impotence under the age of 14 was not changed until 1993.[7] The courts had no discretion on this point of law and therefore never prosecuted young boys on rape charges. Boys were not completely absent from the courts as prisoners, but they were only charged with misdemeanours or as accomplices to the felony charge of rape. The youngest prisoner at trial across both regions was ten years old and was tried for a misdemeanour. He was tried in Gloucestershire along with another boy, aged 15, for an indecent assault on a 12-year-old girl in 1867.[8] The younger boy was acquitted and the slightly older male was sentenced to a period in a reformatory school. Although there was no medical role in this case, the outcome represents a wider societal expectation that young boys were unlikely to perpetrate such offences unless led astray by an older male.

The law on male impotence shaped legal responses to males as both victims and offenders, as the age of 14 came to be an implicit age of consent for boys. In 1880 consent was removed as a defence for cases of alleged indecent assault upon boys under the age of 13. In sodomy cases a boy under 14 could also only be tried as the 'passive' partner, which often involved not being prosecuted at all. As Kim M. Phillips and Barry Reay note, these age-based approaches to the offender and victim in cases of male-male sex can be traced back to the medieval period when '[t]he cultural assumption was that the older partner was the penetrator and the younger the penetrated, even if the age difference was minimal'.[9] Although the law for female complainants operated in two tiers, 14 marked the only male age of sexual consent and the only age at which being the active partner in penetrative sex was thought possible.

Medical witnesses repeatedly disputed these clear-cut legal approaches to male sexual maturity, particularly in cases involving older boys as prisoners. In 1871, for example, the Gloucestershire courts tried a boy aged 13 for attempted carnal knowledge. The prisoner's alleged victim, a girl aged ten, testified that he 'caught hold of me by the neck and pushed me up in the hedge ... He put his thing into mine'.[10] The prisoner pleaded guilty, but, as the law on rape did not acknowledge that boys of his age could commit the crime, he could not be tried for a penetrative assault. The surgeon in this case testified, however, that 'there might be penetration without laceration especially with a boy of the prisoner's size'.[11] This testimony implied that the boy was not at full sexual maturity. As the girl would have been physically small, for the 'prisoner's size' to prevent laceration he must have had underdeveloped or undeveloped genitals. However, this testimony also implied that the boy was theoretically capable of sustaining an erection in order to commit the offence. As the ability to sustain an erection and engage in sexual activity was a marker of puberty, this medical testimony indicated that the prisoner was neither an immature boy nor a fully mature man.

There were some debates about changing the law to reflect medical thought on sexual maturity, but no action came of them. The issue came up so regularly that it made its way to the Home Office in 1890, when a Maryport Clerk to the Justices wrote to the Chief Magistrate that 'two boys aged respectively 9 and 11 appear to have committed a rape upon a child of the age of 5 years, actual penetration, according to the statement of a doctor who has examined the child, having taken place'.[12] The letter also referred to another example from case law 'where Williams J. held that a boy under 14 could not be convicted of carnally knowing and abusing a girl although it was proved he had arrived at puberty'.[13] In this case the Chief Magistrate replied, on behalf of the Secretary of State, confirming that the:

> [E]vidence could be acted on to the extent of a conviction for an 'indecent assault' ... It would probably be thought that a flogging

would be a proper punishment and if the magistrates could be certain that the parents of the boys would give them a severe flogging this might make it unnecessary to bring the case into court which must cause real mischief to the little girls mentioned in your letter.[14]

The youth of the boys, against all medical evidence proving their physical capacity to commit a crime, mitigated their offence in the eyes of the law. The final correspondence in this case confirmed that the 'Petty Sessional Court ... after giving full consideration to the case and ascertaining from the police that the boys have been severely chastised by their parents, have deemed it the most prudent course, adopting your suggestion, not to have the case brought before the court for trial and adjudication'.[15] Medical testimony that proved young boys' capacity to engage in penetrative sex seemingly had little influence on the law or on trial outcomes, although influenced members of the courts sufficiently to compel their correspondence on the matter.

It was widely acknowledged that prisoners above the age of 14 had the potential to commit sexual offences, therefore controversies about the law ebbed in such cases. As capacity was no longer a contested issue at this age, at least in relation to mentally and physically 'normal' males, practitioners of medicine and the law both focused instead on evidence of a crime's perpetration. Most medical testimony related to physical signs on a prisoner, such as venereal disease or marks of a struggle. Medical witnesses spoke little on the subject of prisoners' motives at trial, as this subject was legally irrelevant unless the prisoner sought an insanity defence. However, medical literature addressed the subject of motives and informed medical witnesses' background understanding of sexual offences. General medical understandings of physiology and sexual maturity were crucial for understanding, rather than just proving, the perpetration of sexual crimes. As prisoners for indecent assaults could be any age and those for rape over the age of 14, explanations for sexual offences were also age-based. Age-based concerns fell into two categories: explanations for sexual offences against females of the same age, in which the shifting age of the prisoner was significant; and explanations for offences that involved significant age difference, in which the age of the complainant was also important. These two age-based concerns fell broadly into the categories of 'natural' and 'unnatural' crimes respectively, a distinction that was key to medical thought on age and the sexual offender. The 'natural' and 'unnatural' overlapped with emerging ideas about 'normal' and 'abnormal' acts, but the former carried greater religious and moral weight in relation to sexual behaviour.

Explanations for 'natural' crimes were firmly embedded in wider social ideas about age, social class and masculinity. Medical literature emphasised that the reasons behind 'natural' sexual crimes differed according to a prisoner's age. Boys at puberty were thought to have sexual urges without an acceptable outlet, for example in the form of marriage, a phenomenon that anthropologists and historians have found in a range of contexts and describe as the gap between 'biological' and 'social' maturity.[16] Although officially boys could marry at 14 and girls at 12, few members of any social group married at these ages. Average marital ages had dropped over the centuries preceding the Victorian era, but the mean national marriage age rose again from 24.4 for men and 22.9 for women in the 1850s to 26.7 and 24.4 respectively in the late 1910s.[17] The gap between 'biological' and 'social' puberty was therefore a growing concern of the late-nineteenth and early-twentieth centuries. Although some described this gap as 'adolescence', it did not constitute adolescence in the sense of a middle-class life stage between school and work. Rather, it represented a time when boys had sexual feelings without being in an appropriate social position to have sexual intercourse.

Medical writers also emphasised that boys at puberty would develop sexual urges before the ability to control them; such beliefs explain why boys at puberty were thought 'naturally' likely to pursue their sexual urges or be overtaken by lust. Pubescent boys apparently lacked the mature willpower necessary to control their new sexual feelings. Elizabeth Blackwell commented in her book *The Moral Education of the Young in Relation to Sex* that '[t]he years from 16 to 21 are critical years for youth ... [E]very additional year will enlarge the mental capacity, and may confirm the power of will'.[18] Such comments implied that a boy might be capable of sexual intercourse many years before he had the full 'power of will'. Learning to control sexual urges was also a social issue. Medical texts emphasised that, although sexual desire was harder to manage for young males, families or schools should teach boys to take control of their feelings. Popular medical writer William Acton emphasised that for sexual activity '[p]uberty must not be just dawning; it must be in full vigour; hence the necessity of man's controlling his sexual feelings at an early age'.[19] So-called 'moral education' in the religious aspects of sex and sexuality was deemed crucial for boys at puberty, both through organised moral education (as found in the successful Sunday School movement for middle-class girls and boys) and informal familial guidance in moral matters. In the absence of such moral education, as the middle classes feared was the case in working-class families, boys were more likely to suffer a failure of the 'will'.

The difference between 'biological' and 'social' maturity explained why boys might commit sexual crime at puberty or adolescence, but did not justify it. Self-control was a marker of civilisation for the Victorians and Edwardians, increasingly so over the course of the late-nineteenth century as the failure of self-control came to be a symbol of contemporary concerns about disorder, heredity and class. In the early-twentieth century such ideas were also increasingly evident beyond the medical profession, particularly in literature that emphasised the social implications of individual failures of the will. The eugenics movement found a receptive audience for medical ideas about puberty and adolescence in a range of new scientific and non-scientific professions. In 1912 *Eugenics Review* published an article that emphasised the importance of sex education 'for the sake of the proper use of those functions in the interest of self and society'.[20] Such comments built upon concerns about the puberty 'gap' and placed a new emphasis on the importance of self-control for society as well as for individuals.

The gap between the capacity to engage in sex and the full 'power of will' apparently made puberty a dangerous time for boys, who might find themselves victims of seduction as well as perpetrators of sexual offences. Boys at puberty were likely, it was claimed, either to pursue their passions or to be the victim of blackmail by females taking advantage of their uncontrolled lust. Medical writers in particular emphasised that weak-willed pubescent boys could easily find themselves on the wrong side of the law if they fell victim to the charms of precocious girls. After the 1885 Criminal Law Amendment Act raised the age of sexual consent for females, boys could even find themselves on trial for engaging in ostensibly consensual sexual acts with older girls. Medical writers repeatedly raised concerns about the effects of the 1885 law. *The Lancet* commented that 'the proposed alterations offer increased facilities for the preferring of false charges by precocious females, in the hope of extorting "hush money"'.[21] Surgeon Charles Roberts, who opposed legislative change, made similar claims on the basis of his specialist knowledge of anthropometrics. He observed that 'girls [of 15] who have attained to the physical maturity of boys of from 17 to 19 years of age and to the functional maturity of womanhood will probably prove very troublesome wards of the state, and some disagreeable or unfair litigation and punishment may result from the adoption of the Bill in its present form'.[22] Roberts here emphasised that 'normal' boys were less physically mature than their female counterparts. His work implied that a boy at puberty should not be subject to prosecution for giving in to his 'natural' sexual urges with a girl of the same age who was physically more mature.

For many medical writers and social commentators, irrespective of their political stance on sexual consent, girls at puberty were as much symbols of temptation or blackmail as passive victims. On 10 May 1894, the *Standard* newspaper printed an anonymous letter stating that:

> Raising the age of consent to sixteen was undoubtedly a mistake. Young girls between the ages of fourteen and sixteen have now, it is feared, in an immense number of cases, become sources of income to depraved parents by practising the system of blackmailing. Respectable and innocent lads are led away by designing girls, and induced to break the law.[23]

The reference to 'lads' in this article implied a particular concern about boys at puberty, easily 'led' to commit crimes by engaging in sexual activity with girls of a similar age. A. S. Myrtle, a Justice of the Peace and a Doctor of Medicine, wrote in the same issue of the *Standard* that:

> [Y]ou have dealt so openly with the gross abuses which the Criminal Law Amendment Act of 1885 has given rise to ... I also have sat as a magistrate and on the grand jury since the passing of the Act and have felt in the great majority of the charges brought before me that blackmailing, not justice, was the object of the girl and her friends, and that, in these circumstances, it would have been far better to have said at once 'no bill', than send the case for trial.[24]

The 1885 legislation by no means created these anxieties, but it did exacerbate existing fears by removing consent as a defence in many cases. Although Myrtle was particularly vociferous about his anxieties regarding blackmail, as he cited and supported the idiosyncratic works of Lawson Tait on the subject, his comments reflected the tone of much medical and social literature of the time. During debates some Members of Parliament had even opposed the law on these very grounds, citing the danger of extortion if a man was seduced by 'a precocious, well-advanced, and well-developed girl'.[25]

Pubescent boys, even those who lacked moral guidance and self-control, were generally more sympathetic figures in the eyes of medical practitioners than precocious and pubescent girls. Even in the absence of direct medical testimony on blackmail or puberty, such concerns were widely shared by judges, juries and lawyers in court. Many middle-class men and women had supported a higher age of sexual consent as part of a child protection movement. However, they were also inherently sceptical about the innocence of working-class girls at puberty. The defence counsel in a Gloucestershire case echoed medical ideas about puberty when he mentioned, after it had transpired that the 13-year-old complainant in an indecent assault case had 'passed from childhood into womanhood', that 'girls of that age were subject to strange lies and fancies which so possessed them that they were not accountable for what they did'.[26] Figures 6.1 and 6.2 demonstrate how such anxieties about blackmail and false claims manifested themselves in trial outcomes. The graphs depict alleged heterosexual crimes involving prisoners under the age of 18, who were considered to be most vulnerable to their own uncontrolled urges as well as to seduction and blackmail, before and after the 1885 Criminal Law Amendment; they are stacked at 100 per cent for the purposes of comparison.[27]

Examined side by side, the graphs demonstrate no significant rise in acquittals for the older age group. However, they show that after 1885 juries were more likely to convict prisoners on lesser charges in cases involving complainants over the age of 13. This statistical trend must

Figure 6.1 Verdicts in cases with prisoners under the age of 18, before the 1885 Criminal Law Amendment Act

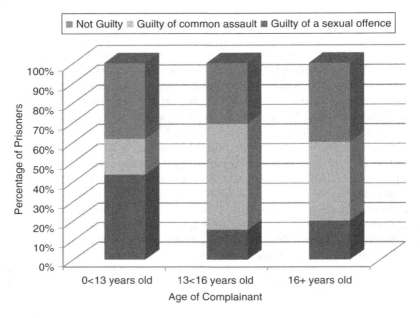

Figure 6.2 Verdicts in cases with prisoners under the age of 18, after the 1885 Criminal Law Amendment Act

be treated with some caution because both graphs draw upon fewer than 100 cases, most of which involved complainants in the youngest age group. However, the small size of the sample for pubescent girls in itself indicates that there was limited sympathy in such cases. Despite significant legal changes, the already small number of complainants aged 13–16 rose little between the two periods – from 11 to 13 – in cases involving prisoners under the age of 18; magistrates seemingly remained reluctant to commit such cases for trial. For those cases that did reach trial, the turn towards verdicts of 'common assault' instead of 'not guilty' indicated jurors' concerns about pubescent boys and girls alike. It allowed for mutual responsibility, without situating either pubescent boys or girls automatically in the role of 'victim' or 'perpetrator' of the act. Changes in the law on sexual consent brought these concerns to the fore when it removed consent as a defence for girls under the age of 16. The 1885 Criminal Law Amendment Act forced jurors to reassess their notions of victimhood, especially as it did not protect boys on the same terms as girls, and led them to use greater discretion by interpreting rather than simply implementing the law on sexual consent.

Judges were also generally lenient with sentencing in cases involving young male perpetrators. When George Bateman, aged 17, pleaded guilty to assaulting a 13-year-old girl in 1893, *The Bristol Mercury and Daily Post* reported that '[t]he Chairman said ... [h]aving regard to his age, he should only pass sentence of one month's imprisonment, with hard labour'.[28] In a Somerset case from 1884 'three lads from 12 to 15 years of age' were convicted of a common assault on a servant girl aged 16 years old.[29] The judge passed a sentence of only 14 days imprisonment out of a possible maximum of 12 months. Although the possibility of this girl's consent was theoretically removed by the 'assault' verdict, this sentence implies that some responsibility was placed upon her. She was older than the prisoners and, according to a local newspaper, 'the defence was that the girl was a consenting party'.[30] The notion that the victim was implicitly blamed for her encouragement in this case, even if it did not constitute consent in law, seems increasingly likely when other cases are considered. In another Somerset case from the same year a 'lad' confessed to an indecent assault in his testimony but was given a reduced charge of common assault and a sentence of only a month, possibly because the complainant was 23 years old and '[t]he defence was that the girl began joking with the prisoner'.[31] The fact that these latter two cases came from the same court in the same year indicates that these decisions might have been influenced by the individual preferences of its judge. However, they also related to broader concerns about 'lads' being led astray. Age difference was a crucial, albeit implicit, backdrop to the light sentencing in these cases as they both involved complainants older than the prisoners.

Medical witnesses, judges and juries demonstrated less sympathy when older males gave in to lust. A combination of training and developmental processes would apparently develop the 'power of will' in respectable men, including the respectable working classes. If men failed to develop self-control after sexual maturity then their actions represented a failure of masculinity. This failure differed from that of boys at puberty, an age at which a lack of self-control represented the failure of moral training for which they could not be blamed entirely. Middle-class tolerance of simple lust as an explanation for crime, which underpinned the sympathy of doctors and jurors for pubescent complainants, therefore faded in direct correlation to the age of prisoners. The lack of sympathy for adult male 'lust' was in part due to changing models of masculinity. A range of contemporary thinkers, from doctors to feminists, increasingly contested the so-called 'double standard' of adult sexual behaviour over the course of the nineteenth century. By the turn of the century middle-class manliness was being redefined under the influence of such forces as medicine and eugenics; there was an increasing expectation that men would show the ability to self-regulate their sexual behaviour. The early 1800s had been a temporary aberration, following the 'sexual revolution' of the seventeenth and eighteenth centuries, as a period in which male sexual freedom was relatively widely accepted after early modern anxieties about uncontrolled male lust.[32] Over the course of the late-nineteenth century, a new interest in race and heredity also came to shape anxieties about male weakness. There was a growing scientific and social emphasis on the ability of the mind to control the body in 'higher' races; in this framework, human weakness was not inevitable, but rather was a marker of primitiveness.

Despite a general lack of sympathy for a failure of self-control, short-term states such as 'lustfulness' and the 'heat of passion' were common medical and social explanations for rape. Such links reinforced a belief that rape was a crime of the 'lower' classes, who lacked the mental faculties or training to control their sexual urges. Prisoners generally denied offences, but the few who pleaded 'guilty' often sought to mitigate their actions by attributing the crime to short-term lust. In 1905, for example, the *Bristol Daily Mercury* reported that:

> Samuel Richard Fishlock (23), labourer, on bail, was indicted on a charge of having committed an indecent assault on Ann Maria Dancey ... He pleaded guilty to a common assault ... Addressing the Court in mitigation of punishment, Mr Cranstoun said the prisoner pushed the woman down in the heat of passion, but there was no serious criminal intent.[33]

The prisoner was sentenced to two months out of a possible maximum of 12, indicating a degree of sympathy with the 'heat of passion' defence. However, in the Middlesex and south-west Quarter Sessions conviction rates were around 6 per cent higher for men aged between 18 and 59 than for prisoners under the age of 18.[34] Although not a significant figure in itself, these men were also much more likely to be convicted on the full charge unless they pleaded guilty to a lesser offence.[35]

Only in cases of insanity were men not responsible for their weakened will. Medical writers emphasised that insane men were more likely to commit crimes such as rape, but that not all rapists were insane. In 1869 William A. Guy reported on 5,598 male convicts and claimed that those convicted for sexual crimes constituted 2.84 per cent of the total sample and 7.5 per cent of insane prisoners.[36] Insanity triggered a failure of willpower, causing offenders to direct 'natural' sexual urges towards victims of indiscriminate age through 'unnatural' force. An 1873 *British Medical Journal* article highlighted that 'in lunatics ... want of self-control may burst out in the most atrocious acts at any moment, even in homicide or rape'.[37] Such medical literature did not pathologise any particular type of offence, but highlighted a general propensity to sexual crime amongst the insane as one of many possible symptoms of a weakened will.

Despite some willingness to consider the insanity defence, a sexual offence in itself was not viewed as constituting *prima facie* evidence of mental instability. Pre-trial statements indicate that the insanity defence was relatively unusual in general. In the entire period 1850–1914, only six prisoners in Middlesex, Gloucestershire, Somerset and Devon were acquitted on the grounds of insanity and none of these trials called upon a trained psychiatrist. Such defences were rare, in part because of the difference between legal and medical definitions of insanity. As Peter Bartlett notes, 'for English law, insanity was not intrinsic to the individual, but was determined by the abilities of the individual in the context of the specific situation'.[38] A pathological sexual deviance or weak will was not the same thing, in the eyes of the law, as an inability to know the difference between right and wrong at a particular moment. Some medical witnesses spoke of a prisoner's general 'imbecility' or 'weak-mindedness', which courts might take into consideration when sentencing, but these labels did not necessarily constitute legal 'insanity'. Just as the sexual instinct in 'imbecile' girls could apparently constitute consent, the uncontrolled sexual drive in 'weak-minded' men constituted agency in law.

These discussions about lust and willpower in relation to 'natural' offences, against females of similar age, focused on pubescent and adult males. The oldest group of prisoners rarely fell into the category of 'natural' offences, as in Middlesex and the South West 82 per cent of alleged sexual crimes involving prisoners over the age of 60 were against complainants under the age of 16. A further 11 per cent of complainants were of unspecified age, leaving only 22 known cases in which so-called 'elderly' men committed offences against women aged 16 or above. These cases were generally rare because men apparently lost their physical strength with old age, sometimes along with their capacity to engage in penetrative sex. They lacked the lustful masculinity that characterised violent, 'real rape' in contemporary eyes, but could still commit offences upon the weak and the young.

Overall, when males committed crimes against females of their own age, there was a widespread social and scientific assumption that offences were sexually motivated. It may seem tautological to refer to sexually-motivated rape or indecent assault, but the existence of an innate link between sexual desire and sexual assaults has been questioned in recent years. Most notably, in the 1970s and 1980s feminist campaigners including Susan Brownmiller began to claim that rape was a crime of power instead of sex.[39] However, medical practitioners in the late-nineteenth and early-twentieth centuries did not understand sexual offences in power terms. Medical writers instead understood sexual crimes in terms of the sexual urges that emerged with masculinity and the imperfect development of self-control, particularly amongst pubescent boys and people of the lower classes. Unlike medical testimony on the female body, medical witnesses in court spoke little of offenders' motives and did not testify directly on these subjects. However, medical theory on the links between lust and self-control shared many characteristics with the approaches of juries, defence lawyers and even prisoners themselves. Medical thought on masculinity was as much social as scientific and drew upon shared middle-class concerns, particularly about the morality of the working classes.

Cases of 'unnatural' crime could not be explained in the same way. 'Unnatural' offences provoked more middle-class anxiety than crimes related to 'natural' sex, as they threatened to destabilise the family. Most so-called 'unnatural' offences involved non-reproductive sex, as was the case with homosexual acts and crimes against the very young, or went against religious doctrine as with incest. Medical and social commentators often proved reluctant to acknowledge that specific, 'unnatural' forms of lust might drive such 'unnatural' offences. In the face of a growing number of 'unnatural' offences coming before the courts, medical explanations desexualised these crimes and provided a range of rational explanations for the misdirection of 'natural' urges. These explanations operated both to deny the possibility of widespread sexual attraction to children and to construct such offences as a problem only of the working classes. Despite what Jeffrey Weeks describes as the growing 'labelling zeal' of nineteenth-century medicine, particularly within sexology, medical witnesses were somewhat reluctant to pathologise sexual offences unless in extreme cases such as 'pederasty'.[40] The more 'unnatural' the crime, the more likely an offender was to be 'monster' rather than 'everyman'.

Despite widespread concerns about 'unnatural' crimes, there was only a limited contemporary belief that such offences resulted from a specific type of sexual urge. Crimes against the very young were either desexualised or understood in terms of general 'lust', and medical practitioners did not automatically pathologise or label the offender. In the construction of 'unnatural' offences, medical men drew less upon models of offender 'types' and more upon a middle-class culture that increasingly feared the working classes and rejected any notion that 'normal' men would sexualise innocent children. There was no widespread unifying term applied to men who committed sexual offences against the young. Newspapers reporting on the Middlesex Sessions used labels such as 'monster', 'beast', 'brute' and 'scoundrel' to describe men convicted of assaults on girls ranging from six years old to adulthood in no consistent pattern.[41]

The 'paedophile' was not yet clearly constructed as a category of offender. Although sexologist Richard von Krafft-Ebing first used the phrase 'paedophilia erotica' in 1886, he claimed that this 'morbid disposition' was only one of many explanations for sexual offences against children that also included 'moral renegades', 'psychico-moral weakness' and 'lasciviousness'.[43] The term also did not appear in the first English translation of his work and was only added to a translation of the tenth edition in 1899.[44] Even translation did not guarantee entry into mainstream English medical theory, as the *British Medical Journal* described *Psychopathia Sexualis* as a 'repulsive' work that had gained popularity 'due rather to the curiosity of the public than to the appreciation of the medical profession'.[45] Differing slightly from Michel Foucault's comparison between the 'sodomite' and the 'homosexual', late-nineteenth century medicine continued to construct sexual offences against children in terms of acts rather than identities.[46]

Perpetrators of offences against the young were not medicalised or treated as a specific type of person. However, generally lustful and weak-willed people were thought to be the most likely to commit all sexual crimes (including those against children and other 'unnatural' offences). Sarah Toulalan has shown that, in the early modern period, men who committed offences against the young were 'characterized as generally immoral, lewd, lustful, and loose-living, notable for [their] debauchery and lack of self-mastery'.[47] Despite the emergence of new theories about sexual deviance in the nineteenth century, particularly from sexology, these older ideas continued to resonate within medicine and wider culture. In the 1870s and 1880s articles in *The Lancet* and *The Times* both referred to 'brutal lust' as a possible motivation for indecent assaults on children.[48] Witnesses echoed the same sentiments at trial, for example in 1865 when George Spencer, aged 21, was accused of indecently assaulting an 11-year-old girl. A witness testified that she saw him assaulting the girl and 'I said to him are you not ashamed of yourself? ... [I]f you are so lustful as that why didn't you go out and cohabit with a woman and not with a child like that?'[49] This language was a continuation of earlier ideas about 'lewd' men and did not differentiate between sexual attraction to children and general 'lust' for women. Such comments did not acknowledge the widespread nature of sexual attraction to children; instead they emphasised that sexual crimes against children only occurred when men's sexual urges were exaggerated and uncontrolled.

The lust that fuelled sexual offences against children was conceptualised as misdirected 'natural' lust, rather than a specific attraction to the young. This misdirection was explained in a range of ways, which served to emphasise that such an act was not 'normal' or typical. One such explanation was grounded in the general relationship between a lack of mental capacity and uncontrolled lustful behaviour. Medical witnesses occasionally sought the insanity defence in cases of alleged assault against the young. In 1880 surgeons in two cases, involving seven- and nine-year-old complainants, described the prisoner as 'exceedingly erratic' and commented that 'I am doubtful of prisoner's sanity'.[50] However, the insanity defence was not a defining characteristic of the 'unnatural' sexual crime and did not follow any pattern along lines such as the age of victims.

More commonly, medical witnesses linked sexual offences against children to weakness of will or imbecility. These states did not constitute an insanity defence, but could explain or mitigate a prisoner's actions. In June 1864, for example, *The Times* reported on a Middlesex case that 'Joseph Coates, aged 15, who bore all the appearance of imbecility, was charged with indecently assaulting a child under three years of age'.[51] When William Martin, aged 24, allegedly committed an offence on a six-year-old boy, surgeon Warwick Charles Steele examined the prisoner and 'found him of weak intellect but capable of understanding what he was doing'.[52] 'Weak intellect' provided some medical explanation for the misdirection of sexual urges, but failed to meet the legal requirement for a successful insanity defence. It was also diagnosed on a case-by-case basis and did not constitute an innate quality of the perpetrators of 'unnatural' offences. However, such comments indicate the broad links between weak intellect and sexual offences that pervaded medical and wider culture. Medical testimony did not construct the sexual offender against children as a psychiatric or personality 'type', but it provided some possible reasons why 'natural' lust might be misdirected towards children and emphasised the abnormality of such behaviours.

Medical writers and witnesses did not deny the extent of sexual offences against the young, but denied the existence of a specific sexual desire for children. In addition to weak intellect or imbecility, they claimed, alcohol fuelled 'unnatural' lust and contributed to the failure of self-control. When David Russell was accused of indecently assaulting 'three little girls' in 1871, *Lloyd's Weekly Newspaper* wrote that '[t]he defence was that ... his intelligence had been weakened to such an extent that he, having had a little whisky to drink, was no longer master of his own actions'.[53] This report drew upon two medical witnesses' testimony for the defence, in which they stated that he had 'drink lust from anxiety ... if he had taken 4 or 5 glasses of whiskey and water I should say the effect would be to confuse him ... would make him a madman' and that 'I believe that the effect of his taking too much stimulant he would lose all control over himself and all moral consciousness'.[54]

Such ideas continued into the early-twentieth century, with the *British Medical Journal* reporting in 1907 that 'chronic alcoholics are very liable to sexual crimes, rape, and indecent assault, also to exhibitionism, assaults on children, and various sexual perversions'.[55] This article made no distinction between the motives for sexual crimes against different victims. Although it noted 'assaults on children' as a special issue of contemporary concern, the *British Medical Journal* did not distinguish between the motives for different sexual offences. Excessive alcohol consumption was both a symptom and a cause of weak will, which could apparently bring on temporary moral insanity and lead to indiscriminate (sexual) crimes. Medical testimony on the links between alcohol and sexual crime sheds light on middle-class concerns about working-class masculinity and on shifting attitudes to gender relations, in part linked to a growing temperance movement that cast women as victims of male alcohol abuse, but reveals little about attitudes to 'unnatural' offences.[56] As Louise Jackson also notes, drinking was also a constituent of 'certain working-class archetypes of manliness' and was thought to have particularly detrimental effects if complainants had suffered from sunstroke in the colonies; such ideas reflected medical ideas about the links between heat and passion.[57]

Lustfulness could not explain all sexual offences against the very young due to the sheer quantity of such cases. Medical writers were somewhat reluctant to explain *all* offences against the very young in terms of lust, which needed to be constructed as abnormal (the result of *extreme* lustfulness, imbecility or alcohol consumption) in order to avoid any recognition of the child as common sexual object. Medical writers and witnesses therefore sought also to identify some non-sexual explanations for sexual offences, including long-rebuked 'superstitions' about curing venereal disease. They explained such crimes as the result of particular social environments and as crimes of limited demographic groups, such as the elderly working class. These explanations not only reinforced the idea that sexual offenders rarely had a *specific* sexual desire for children, but also sought to limit the place of 'lust' in such explanations. They drew upon a range of growing social movements that were anxious about working-class morality and revered childhood innocence.

Beyond the broad-brush explanations of insanity and lustfulness, which could apply to any type of sexual crime, medical approaches to sexual offences differed according to the age of prisoners and complainants. The greatest age gaps between prisoner and complainant marked the most 'unnatural' crimes. At puberty, prisoners' 'unnatural' offences were typically against girls less than a decade younger. In the light of a relatively small age gap and the variable nature of sexual maturity, medical testimony focused not on motives but on physical development in order to assess the relative maturity of the prisoner and complainant. In 1888, when a boy aged 15 was tried for an assault on an eight-year-old girl, the medical witness testified that 'I considered him very well developed in proportion to his years. I should say he is more than 14 years of age'.[58] His focus on the age of 14 related to the law on rape, which required a boy to be aged 14 in order to be tried. The testimony also drew upon new medical concepts of bodily 'norms', as relative rather than absolute. It was medically and legally significant that a boy was well developed 'in proportion to his years', because statistical studies showed that females might develop at a faster rate than males. This testimony made it clear that the prisoner was stronger and more mature than the complainant, a fact that could not be assumed on the basis of age alone; had the medical witness found the complainant to be precocious and the boy immature for his years, the perceived dynamic between complainant and victim would have differed. Advanced physical maturity of this kind also raised the possibility that the prisoner was sexually mature in other ways, for example in terms of sexual knowledge or mental capacity. Although physical in focus, such testimony drew upon broader links between sexual maturity, sexual desire and sexual crime. As the different aspects of puberty were interconnected, even though they did not develop at the same rate, a physically mature boy should have had the emerging capacity to control his urges.

With a wider age gap it was harder to attribute such crimes to an underdeveloped will or to 'natural' sexual urges. Medical writers sought to identify rational explanations why 'normal' adult men might perpetrate such crimes wilfully and in a sane state of mind. They identified a number of possible explanations, using them to avoid any questions about the widespread sexualisation of children. In 1885 *The Lancet*, for example, provided some possible explanations for sexual crimes against children. It commented firstly on the desire to avoid venereal disease and then on 'the hideous superstition ... that connexion with a virgin is a cure for venereal diseases'.[59] Two years later *The Lancet* made similar claims, this time citing the influential German forensic author Johann Ludwig Casper, who stated that the belief that venereal disease was cured by 'coitus with a pure virgin, and most indubitably with a child' was prevalent 'among the lower classes'.[60] By using rational explanations for 'unnatural' sexual acts, of avoiding and curing disease, medical professionals avoided the tricky subject of child sexuality. Only in a third explanation, which related to the 'vicious' desire of old men, did *The Lancet* acknowledge a personality type attracted to children, but it gave far more attention to explanations related to disease.

The so-called 'superstition' of curing venereal disease by sex with a virgin was a widespread trope with a long history.[61] Its persistence into the nineteenth century therefore cannot be explained purely by a continued, genuine belief in the superstition. Instead, it marks a medical and legal willingness to believe that the uneducated working classes embraced such an outdated belief, explained by their apparent reliance on quack literature. Due to the centrality of virginity to such claims, the venereal disease 'superstition' apparently operated in relation to the youngest girls and even infants. Alfred Swaine Taylor, in his textbook of medical jurisprudence, described the belief that sexual intercourse with a female virgin could cure gonorrhoea and syphilis as a 'deplorable, vulgar error, causing this crime to be a frequent one'.[62] Speculation about whether the superstition was 'real' is somewhat futile. Over 30 per cent of medical testimony about prisoners noted the presence of venereal disease, but three quarters of prisoners were never examined by medical practitioners. It is also possible that the high venereal disease rate at the Middlesex Sessions and south-west Quarter Sessions simply corresponded to the prevalence of such diseases in wider society. As Carol Smart has noted in her work on the early-twentieth century, when the myth still persisted, regardless of whether the belief was truly held it 'operated discursively to represent the contact between adult male genitalia and immature female vaginas as completely non-sexual in nature'.[64] By claiming that sexual intercourse with young girls was widely thought to cure venereal disease, writers of medical jurisprudence deliberately constructed 'normal' men who committed such offences as driven by non-sexual motives. The value of the 'superstition' lay in this denial and in its emphasis on the lack of education amongst offenders. Similar discussions endure in recent literature on the 'virgin myth' and HIV/AIDS in Africa from the late-twentieth century, which operate similarly to construct the offence in terms of ignorance or cultural tradition instead of sex.[65]

The tone of medical literature shifted again when the elderly committed sexual offences against the very young, an age gap that indicated a particularly 'unnatural' crime. Newspapers reporting on the Middlesex Sessions and south-west Quarter Sessions described prisoners as 'elderly' above the age of 60, meaning that such cases involved age differences of around 50 years between prisoner and complainant. *The Lancet*, in addition to the two motives linked to venereal disease, referred to a third and final reason for offences against children: 'a depraved taste on the part of elderly men exhibited in a vicious demand for young girls'.[66] Such a claim marked a slight shift from the early modern period in which, as Garthine Walker notes, it had been thought that aging men 'lack[ed] the overwhelming physiological and mental passions that motivated rape'.[67] For the Victorians, the declining 'mental' faculties of the elderly marked not a retreat from lust but an increasing inability to control it. *The Lancet* was not alone in linking sexual desire for children to elderly offenders rather than to all men because it was thought that, as part of the natural life cycle, male willpower weakened again in old age. As the faculties diminished, rationality in relation to choice of sexual partner was thought to diminish. Elderly men also apparently lost the physical strength to commit rape on adult women and often became impotent, which (rather than declining 'passions') may explain the particular link between the elderly and indecent assaults upon the young.

Stereotypes about the elderly being prone to a 'vicious' demand for the young were widely embedded in late nineteenth-century and early twentieth-century society. However, it seems unlikely that practical experience would have led many general practitioners or police surgeons truly to believe this to be the case. 52 per cent of the accused across all Quarter Sessions and the Middlesex Sessions were under the age of 30, and only 7 per cent were over 60 years old. Rather than truly believing that the sexual 'taste' for children was limited to the elderly, in the face of alternative empirical evidence, it seems that medical practitioners deliberately utilised a prevalent Victorian stereotype to deny that young girls and boys would be viewed in sexual terms by 'normal' men. Working-class elderly men were in a demographic minority; limiting sexual explanations for 'unnatural' offences to this social group therefore also limited the recognition of child sexuality.

In general, the courts showed some sympathy for the actions of the elderly. While adult men were expected to act rationally and with control, the elderly lost some willpower and control over their sexual functions as a natural result of aging. Juries often recommended elderly men for mercy in sentencing on the basis of their age. Some judges followed these recommendations, as in a Gloucestershire case from 1864 when '*George Clement*, an old man 73 years of age, was indicted for committing a criminal assault upon a little girl ... the Chairman, in passing sentence of nine months' imprisonment, said that but for his age of the punishment would have been much more severe'.[68] However, judges were less sympathetic when an elderly man was a repeat offender. In another Gloucestershire case from 1884, according to *The Bristol Mercury and Daily Post*:

> An old man named *Isaac Stone*, a hawker, was charged with indecently assaulting Catharine Bowby, aged six years ... He was found guilty, but was recommended to mercy on account of his age. The Chairman sentenced him to 12 months' hard labour, as he had been twice before convicted of similar offences.[69]

Repeat crimes of this kind indicated a 'vicious demand', in *The Lancet*'s words, rather than an isolated lapse because of the prisoner's declining mental faculties. Although the 'rapist' and the 'paedophile' both emerged as labels in the nineteenth century, it was the repeat offender who was most widely accepted to be a 'type' of person because of his crimes.

In the early-twentieth century the nascent discipline of psychiatry linked 'senile dementia' to acts of indecency and became increasingly sympathetic to such acts. Rather than viewing them as 'vicious' sexual desire for the young, writings on senile insanity and dementia compared the very old to the very young; as men became more childlike in their mental functions, sexual acts lost their meaning. Medical writers on dementia emphasised that, like the young, such men should be protected from their own uncontrolled urges rather than blamed for them. In 1907, for example, the *British Medical Journal* published an 'Address on the Treatment of Incipient and Borderland Cases of Insanity in General Practice' that stated 'the early symptoms of the old man nearing his dotage are well known to you ... You must also safeguard the patient against indecent exposure of the person and other acts, the outcome of forgetfulness or senile imbecility'.[70] With a growing interest in dementia there was a turn away from deliberate sexual 'demand' as an explanation for the behaviour of the elderly, leaving very few cases of offences against children that could be explained in such a way. In the 1920s and 1930s, medical writers wrote of elderly regression and 'senile dementia' as a cause of sexual offences against the young.[71] Only occasionally did writers such as the Swiss Professor of Psychiatry Auguste Forel, whose work an English surgeon translated and expanded, indicate that men might have sexual desire for children throughout adult life.[72] In his 1931 work even Forel, however, emphasised that such a perversion was rare and that 'many sexual assaults committed on children are simply the effect of senile dementia, or abuse of children to satisfy an otherwise normal sexual appetite'.[73] There was a continued emphasis on the elderly as a social group prone to assaulting the young, although the growing emphasis on dementia de-sexualised such acts in the early-twentieth century.

The social and medical stereotype of elderly perpetrators related to homosexual, as well as to heterosexual, acts. In 1882, when a 13-year-old boy accused a 72-year-old man of indecent assault, the prisoner apparently stated to a policeman that 'I am very fond of boys ... you know what old men are'.[74] The 'you know' in this statement indicates that such stereotypes about 'old men' were prevalent. However, sexual offences against young boys were particularly difficult to explain. As Louise Jackson notes, 'the assault of boy children was doubly "unnatural" because of the combined stigma of gender and age'.[75] Due to this 'doubly unnatural' aspect, which made the crime particularly difficult to explain in terms of misdirected 'natural' lust or disease prevention, practitioners of sexual forensics acknowledged a personality 'type' with specific sexual desire for male children: the pederast.

Alfred Swaine Taylor first referred to 'pederastia' in the ninth edition of his *A Manual of Medical Jurisprudence*, printed in 1874, as a 'new term' invented by 'continental medical jurists'.[76] He defined the term as referring to cases 'in which boys at about the age of puberty are made the victims of the depraved passions of a certain class of men'.[77] This definition of 'pederasty' drew upon contemporary concerns about the corruption of young and pubescent boys, which was perhaps even the driver for the 1885 Labouchère amendment that criminalised 'gross indecency' between males.[78] The 'pederast' was seemingly a more influential construction than the 'homosexual' or the 'paedophile'; its rare 'doubly unnatural' dimension and perceived links with elderly perpetrators prevented any recognition that this personality type was common. The belief that only a certain type of person committed offences against young boys soon took hold and was reasonably widespread in the UK by the turn of the century. Despite the taboo nature of this topic, in the early decades of the twentieth century some writers on the fringes of English culture – such as James Joyce in Ireland – even dealt with the stereotype of the 'old' man with an unhealthy interest in young boys in their work.[79]

'Doubly unnatural' crimes, in which men committed offences on boys during childhood or puberty, sometimes did not come to light at all. The families of complainants often preferred to deal with such cases informally in order to avoid the social stigma associated with them. In 1914, for example, a Gloucestershire police sergeant deposed that:

> I called upon Mr Edward Pick of the Woodlands, Stinchcombe, who informed me that the above named man [Thomas] went to lodge with him in the early part of the year 1909. He had been there about 9 months when his son Albert, then 11 years of age, made a complaint to him of having been indecently assaulted by Thomas, in consequence of which he took him to Dr Dale Roberts of Dursley, who examined him, and found the lad's person was very red & sore from abuse. Mr Pick never reported the matter, but ordered Thomas to leave his premises at once, which he did.[80]

The decision not to prosecute in this case may have been a consequence of the complainant being male and the crime in question being 'attempted buggery'. This case was only reported because it later reached trial by other means, but raises the possibility that similar cases may have gone undetected. It provides a partial explanation for the low proportion of cases at trial involving male victims, both with and without medical evidence. Although the physician did not actively encourage informal sanctions, his evidence informed the father's decision to evict the suspected offender from his premises rather than prosecute.

Similar trends can be identified in the context of incest cases, which were also 'doubly unnatural' and were more likely to be dealt with informally than other sexual offences. In one notable Devon case from 1871, an 11-year-old girl was apparently 'carnally known' by her brother who was ten years older than her.[81] Both being immediate blood relatives and children of the same mother, who may have feared the shame that such a case would bring upon the family, a charge was not immediately pursued.[82] Instead, the mother testified that:

> I took her to Cullompton to Mr Blanchard a Chemist, who gave me some medicine for her, which I gave her. I saw some marks of corruption on her linen before I took her to Mr Blanchard's. I told the prisoner, who is my son, of it and he denied it. I don't remember what I said to him. I told him what the child told me. I said 'Oh Harry what have you done, you've injured the girl'. He denied it. I did not like to say anything about it to anyone. I told my son he had given his sister the disease ... I sent her out to Mrs Pearce to live afterwards. I thought she was then cured.[83]

The chemist treated the girl but did not examine her or provide any support for a criminal case. The case eventually came to police attention in a complex way. A local general practitioner treated the son of Mrs Pearce, who was seemingly no relative of the complainant or prisoner, for gonorrhoea. This medical practitioner then examined the girl and found that she also still had gonorrhoea, which brought the whole case to light. Although it is difficult to know the extent to which cases were *not* brought before courts, trial reports indicate that parents of complainants pursued informal sanctions most commonly when cases involved pederasty or close family incest. In avoiding bringing such cases to light, members of the public implicitly reinforced a general societal belief that those crimes were rare.

The absence of these cases from trial fuelled their construction as 'abnormal' offences and meant that medical practitioners rarely reached the stand in order to testify about the motives behind them. There was evidently a general reluctance to take the 'doubly unnatural' offences of incest and pederasty to trial, but it is easy to overstate what Jeffrey Weeks refers to as a nineteenth-century 'incest taboo'.[84] *The Times* and London-based newspapers reported on Middlesex Sessions cases involving incest at the same rate as other cases.[85] Anthony S. Wohl argues that *The Lancet* showed 'typical ... Victorian reticence on the subject' of incest, but medical journals did address the subject in order to emphasise its rare and socially-contingent nature.[86] In 1868 the *British Medical Journal*, for example, published the speech of a surgeon and previous London Medical Officer of Health linking unsuitable working-class accommodation to 'incest ... filthy and degraded habits'.[87] Such links between incest and one-room housing were common in late-Victorian society and represented growing concerns about the immorality of urban environments.[88] Such commentaries made no distinction between the causes of incestuous offences against children and between siblings, connecting them all to overfamiliarity caused by living arrangements. Although the focus of such discussions was on the city, they were as much about social hierarchy as urban environment. Contemporary commentators described similar issues of incest and immorality among agricultural workers and, as parts of the South West were also undergoing partial urbanisation, other provincial working-class communities.[89]

'Unnatural' sexual offences were often grouped with other sexual immorality of the working classes. Most late-Victorian medical practitioners accepted that 'normal' men from the labouring classes, although not 'normal' respectable men, could sexually assault children. Crimes against the young were either attributed to the same 'lustfulness' and weakness of will that caused all sexual offences or, in order to explain their particular prevalence, to factors unrelated to sexual desires such as curing disease. This approach supported mainstream medical and middle-class frameworks of thought, within which most medical witnesses operated, which emphasised the relative asexuality of the very young in both social and developmental terms (even bearing in mind the changes that took place over the course of childhood). Medical writers only constructed 'doubly unnatural' offences, which could not be explained easily in the same terms, as limited either to particularly immoral environments or to small demographic groups such as the elderly. Overall, medical practitioners showed a general aversion to categorising and labelling sexual deviance as pathological unless in exceptional cases. Physiologists, writers of medical jurisprudence and surgeons rarely worked in the field of sexology and were as much members of general middle-class communities as of specialist medical communities.

It was only in the late-twentieth century that paedophilia became a widely recognised category due to a combination of the greater visibility of such crimes, the growing influence of sexology and psychology, and the women's rights movement's emphasis that 'child sexual abuse' occurred widely within the home. As 'child sexual abuse' was reframed as a coherent category linked to familial abuse, rather than to strangers, the crime became again 'doubly unnatural' and potentially pathological.[90] Such crimes had previously been the least likely to come to attention and evidence of their prevalence forced a recognition that such crimes could not be explained by general 'lust' or rational acts of disease prevention. Some alternative explanations became less viable, with public health reforms removing working-class slum housing as an explanation for incest, and pathological deviance became an increasingly convincing framework for understanding crime. As paedophilia became widely accepted as a concept it also moved outside of the medical framework and into general society, within which increasing concerns have been articulated about the 'sexualisation' of children. In the late-nineteenth century and early-twentieth century, in contrast, many medical writers and members of general society ignored the possibility of child sexualisation completely.

Case Analysis: Offenders, 1858

The Case: Due to the range of possible explanations for sexual crime, each trial reflected its own particular relationships of age and gender. This case, although not representative of all constructions of the offender and his motives, provides insights into one such set of relations. It looks at the desexualisation of offences against the very young using the venereal disease 'superstition' and includes rare medical testimony about the perceived motives of the prisoner. The case therefore has some general value, beyond a better understanding of the disease 'superstition'; it indicates a degree of overlap between ideas about the sexual offender that medical witnesses brought with them – often implicitly – into the courts and those found in more extensive written discussions about sex, crime and deviance.

The Prisoner: George James, aged 19, Labourer.
The Complainant: Elizabeth Jane Brown, aged two.
The Complaint: Indecent assault (consent no defence).
The Pre-Trial Statement: Depositions taken 11th February 1858, Thames Police Court.[91]

Bridget, wife of William, Brown: I live at 11 Christian Street, Saint George with my husband who is a poulterer. I have an infant child (aged 2 years) named Elizabeth Jane Brown; the prisoner lives with my daughter in my house. They are not married and on Monday evening the 1st instant I gave [the prisoner] my little girl Elizabeth Jane. It was between 7 and 8 o'clock. He had the child for 2 hours. I was in the next room during that time attending upon my daughter who was confined on the Monday evening. On the Tuesday afternoon I washed the baby and found her very sore. She cried very much and a discharge came from her private parts. I took her to Mr Comley the Surgeon (now present). When I took my child from the prisoner on the Monday evening I found her crying.

Sarah Cole: 9 Sidney Street, Mile End, [my husband is] a poulterer. sister-in-law to the last witness. I remember the Monday evening in question. I was in the room with the last witness and I heard the child Elizabeth Jane Brown scream in the next room. I called to prisoner in the next room and told him to stay, not to go out.

John Comley: I am a Member of the Royal College of Surgeons of England and live at 71 High Street, Whitechapel. On Saturday evening … at 12 o'clock I saw the child Elizabeth Jane Brown at the Leman Street Police Station. The first witness was with it. I examined the child and found a discharge from the private parts which I believe to be gonorrhoe [sic]. The parts were much inflamed and a little open and [there was] a very large discharge. The sides of the both legs were also inflamed. This may have been occasioned from contagion with a person suffering from gonorrhoe [sic]. The prurient matter ... might have been communicated otherwise, than by the genital organs of a man. On the same evening I examined the prisoner and found him suffering from that complaint. It is a common notion among a certain class of individuals that a person suffering from gonorrhoe [sic] may be cured by having connection with a female who is not suffering from that disease.

Police: ... He said 'I am innocent of the charge'. Shortly afterwards he said 'the father of the child might have done it', at at [sic] the station he said 'how do you know ... that the mother might not have communicated it?'

The Verdict: Guilty of indecent assault, 2 years hard labour (out of a maximum possible of 2 years).

Significance for Sexual Forensics: There is no representative case of medical attitudes to 'the' sexual offender. However, this trial provides an example of how one stereotype (the venereal disease 'superstition') operated in court and thereby illustrates some of the moral dimensions of medical thought on the offender. This case was notable for the youth of the victim. An assault on an infant, aged two, could not be explained away as misdirected 'natural' and 'normal' sexual urges. The medical witness therefore instead sought to explain the prisoner's actions through the famed venereal disease 'superstition'. He testified that 'it is a common notion among a certain class of individuals that a person suffering from gonorrhoe [sic] may be cured by having connection with a female who is not suffering from that disease'. This emphasis on the 'certain class' of complainant implied the lower social standing of people who believed such a superstition and operated to associate such crimes with ignorance and social class, rather than with sexual desire for children. This medical witness was the only practitioner to refer to the 'superstition' on the stand, in part because the courts rarely solicited medical opinion on motives, but was certainly not alone in such beliefs. Other medical witnesses may therefore have drawn more implicitly upon such theories, which were widely articulated within medical literature, even though testimony on the 'superstition' was rare in court.

The case was tried and convicted as 'indecent assault' rather than 'carnal knowledge', despite medical evidence about the transmission of gonorrhoea. The magistrate may have pursued this charge for its higher conviction rate or, in line with age-based ideas about sexual crime, may have thought a rape to be impossible upon a child so young. The medical witnesses supported the possibility that the disease might have been caught without penetrative sex, noting that it 'might have been communicated otherwise, than by the genital organs of a man'. As the medical witness believed that the 'contagion' came from a man with gonorrhoea, this statement indicated a belief that there was potential for transmission by genital contact, hands or cloth without rape. Whatever the exact mode of infection, the 'superstition' motive desexualised the offence, while the transmission of disease and 'pollution' of a young girl aggravated it. The prisoner was convicted on a lesser charge but given the maximum sentence. *The Morning Post* referred to this venereal disease when it reported that the case had 'most distressing consequences'.[92] The 'superstition' explanation removed a sexual motivation, but did not mitigate an offence.

Notes

1. Steven Angelides, 'The Emergence of the Paedophile in the Late Twentieth Century', *Historical Studies* 36 (2005), 272–95, p. 273.
2. Angelides, 'The Emergence of the Paedophile', p. 273.
3. In the late-twentieth century paedophilia and child sexual abuse came to be inextricably woven, meaning that there was little conception of a paedophile who would not offend.
4. Garthine Walker, 'Everyman or a Monster? The Rapist in Early Modern England, c. 1600–1750', *History Workshop Journal*, 76 (2013), 5–31.
5. Francis Ogston, *Lectures on Medical Jurisprudence* (London: J. & A. Churchill, 1878), p. 127.
6. John Abercrombie, *The Student's Guide to Medical Jurisprudence* (London: J. & A. Churchill, 1885), p. 353. See also Alfred Swaine Taylor, *A Manual of Medical Jurisprudence*, 8th edn (London: J. & A. Churchill, 1866 [1844]), p. 571.
7. This presumption was not removed from law until the *Sexual Offences Act* 1993; Michael Jefferson, *Criminal Law*, 8th edn (Harlow: Pearson Longman, 2007 [1992]), p. 572.
8. Gloucester, Gloucestershire Archives (GA), Pre-Trial Statements, George Morgan and John Nelmes tried at the Gloucestershire Quarter Sessions on 4 July 1867 for indecent assault, Q/SD/2/1867.
9. Kim M. Phillips and Barry Reay, *Sex before Sexuality: A Premodern History* (Cambridge: Polity Press, 2011), p. 71.
10. Gloucester, GA, Pre-Trial Statements, John Thomas Pearson tried at the Gloucestershire Quarter Sessions on 18 October 1871 for carnal knowledge, Q/SD/2/1871.
11. Gloucester, GA, Pre-Trial Statements, Pearson, Q/SD/2/1871.
12. Kew, National Archives (NA), Rape by Boys aged 9 and 11, July–August 1890, HO 45/9721/A51769.
13. Kew, NA, Rape by Boys, HO 45/9721/A51769.
14. Kew, NA, Rape by Boys, HO 45/9721/A51769.

15. Kew, NA, Rape by Boys, HO 45/9721/A51769.
16. On the so-called 'puberty gap' or the difference between 'biological adolescence' and 'social adolescence' see Alice Schlegel, 'The Cultural Management of Adolescent Sexuality' in *Sexual Nature, Sexual Culture*, ed. Paul R. Abramson and Steven D. Pinkerton (Chicago; London: University of Chicago Press, 1995), 177–94, p. 178; and Helen King, *The Disease of Virgins: Green Sickness, Chlorosis and the Problems of Puberty* (London; New York: Routledge, 2004), pp. 87–88.
17. Barry Reay, *Microhistories: Demography, Society, and Culture in Rural England, 1800–1930* (Cambridge: Cambridge University Press, 1996), p. 41; and Michael Anderson, 'The Social Implications of Demographic Change' in *The Cambridge Social History of Britain 1750–1950*, ed. F. M. L. Thompson, vol. 2 (Cambridge: Cambridge University Press, 1990), 1–70, p. 32.
18. Elizabeth Blackwell, *The Moral Education of the Young in Relation to Sex*, 6th edn (London: Hatchards, 1882 [1878]), p. 62.
19. William Acton, *The Functions and Disorders of the Reproductive Organs in Youth, in Adult Age, and in Advanced Life, Considered in their Physiological, Social and Psychological Relations* (London: John Churchill, 1857), p. 14.
20. J. A. Russell, 'The Eugenic Appeal in Moral Education', *The Eugenics Review*, 4 (1912–13), p. 136.
21. F. W. Lowndes, 'The Criminal Law Amendment Bill', *The Lancet*, 1 August 1885, 221–22, p. 222.
22. Charles Roberts, 'The Physical Maturity of Women', *The Lancet*, 25 July 1885, 149–50, p. 149.
23. 'The Criminal Law Amendment Act', *Standard*, 10 May 1894, 6, p. 6. For similar comments see 'Common Sense', *The Times*, 3 November 1887, 13, p. 13.
24. 'The Criminal Law Amendment Act', *Standard*, 10 May 1894, 6, p. 6.
25. HC Deb. (3rd series), 3 August 1885, vol. 300, c. 898.
26. 'Thornbury: Police Court', *The Bristol Mercury and Daily Post*, 10 June 1885, 7, p. 7.
27. Charles Roberts highlighted concerns about blackmail because girls of fifteen 'have attained to the physical maturity of boys of from seventeen to nineteen years of age'; Roberts, 'The Physical Maturity of Women', p. 149. A sample of cases involving prisoners under the age of 18 is therefore appropriate in order to consider links between age difference and blackmail concerns.
28. 'Gloucestershire Quarter Session', *The Bristol Mercury and Daily Post*, 7 January 1893, 3, p. 3; see also Taunton, Somerset Heritage Centre (SHC), Pre-Trial Statements, Samuel Rendell tried at the Somerset Quarter Sessions on 2 July 1884 for assault with intent, Q/SR/736.
29. 'Somerset Quarter Sessions', *The Bristol Mercury and Daily Post*, 4 January 1884, 6, p. 6.
30. 'Somerset Quarter Sessions', *The Bristol Mercury and Daily Post*, 4 January 1884, 6, p. 6.
31. 'Somerset Midsummer Session: Assault', *The Bristol Mercury and Daily Post*, 3 July 1884, 3, p. 3.
32. Faramerz Dabhoiwala, *The Origins of Sex: A History of the First Sexual Revolution* (Oxford: Oxford University Press, 2012).
33. 'Gloucestershire Quarter Sessions', *Bristol Daily Mercury*, 19 October 1905, 3, p. 3.

34. In cases involving prisoners under the age of 18, 125 were found guilty out of 266 (47%). For those aged 18–59, 995 were found guilty out of 1866 (53%).
35. In cases involving prisoners under the age of 18, of the 125 found guilty: 43 were convicted on lesser charges (34 % of guilty verdicts). For those aged 18–59, of the 995 found guilty: 175 were convicted on lesser charges (18 % of guilty verdicts).
36. W. A. Guy, 'On Insanity and Crime: and on the Plea of Insanity in Criminal Cases', *Journal of the Statistical Society of London* 32 (1869), 159–91, pp. 170–71.
37. C. B. Radcliffe, 'Croonian Lectures on Mind, Brain, and Spinal Cord, in Certain Morbid Conditions', *British Medical Journal (BMJ)*, 5 April 1873, 359–63, p. 359.
38. Peter Bartlett, 'Legal Madness in the Nineteenth Century', *Social History of Medicine* 14 (2001), 107–31, p. 109.
39. Susan Brownmiller, *Against Our Will: Men, Women and Rape* (Harmondsworth: Penguin, 1977 [1975]).
40. Jeffrey Weeks, *Sex, Politics, and Society: The Regulation of Sexuality since 1800*, 3rd edn (London: Longman, 2012 [1981]), p. 29.
41. 'Middlesex Sessions', *Reynolds's Newspaper*, 29 May 1870; 'Middlesex Sessions', *Reynolds's Newspaper*, 13 June 1886; 'Middlesex Sessions', *Pall Mall Gazette*, 10 June 1870; 'Trials at the Middlesex Sessions', *Lloyd's Weekly Newspaper*, 16 May 1886; 'General News', *Illustrated Police News*, 13 March 1886.
42. Weeks, *Sex, Politics, and Society*, p. 29.
43. Richard Von Krafft-Ebing, *Psychopathia Sexualis: with especial reference to antipathic sexual instinct: a medico-forensic study*, trans. by C.G. Chaddock, 10th edn (London: Rebman, 1899), p. 525.
44. Krafft-Ebing, *Psychopathia Sexualis*, 10th edn, p. 525.
45. 'Sexual Psychology and Pathology', *BMJ*, 8 February 1902, p. 339; 'Obituary: Freiherr Von Krafft-Ebing, M.D.', *BMJ*, 3 January 1903, 53, p. 53.
46. Michel Foucault, *The History of Sexuality, Vol. 1. An Introduction*, trans. Richard Hurley (New York: Pantheon, 1978), p. 43.
47. Sarah Toulalan, ' "Is He a Licentious Lewd Sort of a Person?": Constructing the Child Rapist in Early Modern England', *Journal of the History of Sexuality* 23 (2014), 21–52, p. 22.
48. F. W. Lowndes, 'The Criminal Law Amendment Bill', *The Lancet*, 1 August 1885, p. 222; 'Middlesex Sessions', *The Times*, 8 Jun 1870, p. 11.
49. London, London Metropolitan Archives (LMA), Pre-Trial Statements, George Spencer tried at the Middlesex Sessions on 4 July 1865 for indecent assault, MJ/SP/E/1865/013.
50. London, LMA, Pre-Trial Statements, Charles Herbert tried at the Middlesex Sessions on 7 October 1880 for indecent assault, MJ/SP/E/1880/035; London, LMA, Pre-Trial Statements, Philip Powys tried at the Middlesex Sessions on 7 April 1880 for indecent assault and assault, MJ/SP/E/1880/012.
51. 'Middlesex Sessions, June 6', *The Times*, 7 June 1864, 13, p. 13.
52. London, LMA, Pre-Trial Statements, William Martin tried at the Middlesex Sessions on 8 August 1888 for indecent assault, MJ/SPE/1888/045.
53. 'Saturday's Law and Police', *Lloyd's Weekly Newspaper*, 30 July 1871, 8, p. 8.
54. London, LMA, Pre-Trial Statements, David Russell tried at the Middlesex Sessions on 9 August 1871 for indecent assault, MJ/SP/E/1871/015.

55. 'Seventy-Fifth Annual Meeting of the British Medical Association', *BMJ*, 28 September 1907, 795–826, p. 795.
56. Lucy Bland, ' "Purifying" the Public World: Feminist Vigilantes in Late Victorian England, *Women's History Review* 1 (1992), 397–412.
57. Louise A. Jackson, *Child Sexual Abuse in Victorian England* (London: Routledge, 2000), pp. 118–9.
58. Gloucester, GA, Pre-Trial Statements, Joseph Henry Roberts tried at the Gloucestershire Quarter Sessions on 5 July 1888 for indecent assault, Q/SD/2/1888.
59. 'The Criminal Law Amendment Bill', *The Lancet*, 8 August 1885, 252, p. 252.
60. Frederick W. Lowndes, 'Venereal Diseases in Girls of Tender Age', *The Lancet*, 22 January 1887, 168–69, p. 169.
61. On the early modern manifestations of this 'superstition' see Toulalan, 'Constructing the Child Rapist in Early Modern England', pp. 41–45.
62. Taylor, *Manual*, 11th edn (London: J&A Churchill, 1886), p. 690.
63. 'The Criminal Law Amendment Bill', *Lancet*, 8 August 1885, 252, p. 252; for further reading on W. T. Stead's articles see Deborah Gorham, 'The "Maiden Tribute of Modern Babylon" Re-Examined: Child Prostitution and the Idea of Childhood in Late-Victorian England', *Victorian Studies*, 21 (1978), 353–379.
64. Carol Smart, 'A History of Ambivalence and Conflict in the Discursive Construction of the "Child Victim" of Sexual Abuse', *Social and Legal Studies* 8 (1999), 391–409, p. 397.
65. For example see Suzanne Leclerc-Madlala, 'On the Virgin Cleansing Myth: Gendered Bodies, Aids and Ethnomedicine', *African Journal of AIDS Research*, 1 (2002), 87–95.
66. 'The Criminal Law Amendment Bill', *The Lancet*, 8 August 1885, 252, p. 252.
67. Garthine Walker, 'Everyman or a Monster? The Rapist in Early Modern England, c. 1600–1750', *History Workshop Journal* 76 (2013), 5–31, p. 15.
68. 'Somerset Quarter Session', *The Bristol Mercury*, 2 July 1864, 6, p. 6.
69. 'Gloucestershire Quarter Session: Indecent Assault', *The Bristol Mercury and Daily Post*, 3 July 1884, 3, p. 3.
70. 'An Address on the Treatment of Incipient and Borderland Cases of Insanity in General Practice', *BMJ*, 9 March 1907, 546–49.
71. For example, see William Norwood East, 'The Interpretation of some Sexual Offences', *Journal of Mental Science*, 71 (1925), 410–24, p. 421. Thanks go to Janet Weston for recommending this reading on old age and sexual offences.
72. Auguste Forel, *The Sexual Question*, trans. C. F. Marshall (Brooklyn, New York: Physicians and Surgeons Book Co., 1931 [1906]), p. 254.
73. Forel, *The Sexual Question*, p. 254.
74. London, LMA, Pre-Trial Statements, George Henry Winkfield tried at the Middlesex Sessions on 16 October 1882 for indecent assault, MJ/SP/E/1882/043.
75. Jackson, *Child Sexual Abuse*, p. 102.
76. Taylor, *Manual*, 9th edn (London: J. & A. Churchill, 1874 [1844]), p. 677.
77. Taylor, *Manual*, 9th edn, p. 677.
78. Joseph Bristow, 'Wilde, Dorian Gray, and Gross Indecency' in *Sexual Sameness: Textual Differences in Lesbian and Gay Writing*, ed. Joseph Bristow (London: Routledge, 1992), 44–63, p. 49.
79. See 'An Encounter' in *Dubliners* (1914).

Offenders: Lust and Labels 189

80. Gloucester, GA, Pre-Trial Statements, Walter Thomas tried at the Gloucestershire Quarter Sessions on 21 October 1914 for attempted buggery, Q/SD/2/1914.
81. Exeter, Devon Record Office (DRO), Pre-Trial Statements, Henry Clarke tried at the Devon Quarter Sessions on 19 October 1871 for carnal knowledge, QS/B/1871/Michaelmas.
82. On the 'exposure, shame, cowardliness and humiliation associated with incest' see Jackson, *Child Sexual Abuse*, p. 121.
83. Exeter, DRO, Pre-Trial Statements, Clarke, QS/B/1871/Michaelmas.
84. Weeks, *Sex, Politics, and Society*, p. 31.
85. Based on an analysis of cases reported from the Middlesex Sessions in *The Times*; *Lloyd's Weekly Newspaper*; *Illustrated Police News*; *Reynolds's News*; *Daily News*; *Pall Mall Gazette*.
86. A. S. Wohl, 'Sex and the Single Room: Incest among the Victorian Working Classes' in *The Victorian Family: Structure and Stresses*, ed. A.S. Wohl (London: Croom Helm, 1978), p. 201.
87. W. Rendle, 'The Residential Conditions of the Lower Classes of Inhabitants of this Kingdom, with reference to Health and Complementary Conditions', *BMJ*, 22 August 1868, 207, p. 207.
88. 'The Criminal Law Amendment Bill', *The Lancet*, 8 August 1885, 252, p. 252; Weeks, *Sex, Politics, and Society*, p. 31.
89. The *First Report of Her Majesty's Commissioners for Inquiring into the Housing of the Working Classes* (London: HMSO, 1885) dealt with incest, immorality and overcrowding in provincial cities such as Bristol, and agricultural cottages in locations such as Somerset and Dorset, as well as cities more commonly associated with these issues such as London.
90. Although there are debates about the extent to which paedophilia can be considered a form of mental health issue, it is listed within the *DSM* as a 'paraphilia'.
91. London, London Metropolitan Archives (LMA), Pre-Trial Statements, Elizabeth Jane Brown tried at the Middlesex Sessions on 17 February 1858 for assault with intent, MJ/SP/E/1858/004. Full transcript, except for occasional sentences with illegible words. Some punctuation added for clarity.
92. 'Middlesex Sessions.— February. 17', *The Morning Post*, 18 February 1858, 7, p. 7.

Conclusions: Medicine, Morality and the Law

The nineteenth century was a period of important change in medical knowledge and practice. Forensic medicine developed as a profession in the Victorian period and established an increasingly coherent body of knowledge, which included the legal aspects of sexual crime. Science carried some weight on the witness stand, where practitioners of sexual forensics testified on the legal implications of a complainant's physical and mental signs in the aftermath of an alleged sexual crime. Medical witnesses needed to consider all possible meanings of a sign in order to identify whether it held legal relevance. Sexual forensics therefore dealt with subjects that were outside the remit of most witness testimony, ranging from a complainant's general physical maturity to her (and sometimes his) chastity.

Sexual forensics was grounded in knowledge about sexual development and particularly in ideas about the healthy (and, by extension, abused or unhealthy) female body at different life stages. The changes that came at puberty were of particular interest because they altered the legal significance of important bodily signs, such as bleeding, and apparently brought with them the capacity for consent or resistance. It is therefore significant that, around the time that forensic medicine emerged as a profession, the female pubescent body was being redefined in physiology. Statistical studies of sexual development contributed to a new meaning of 'normal', as equivalent to 'typical' and as a relative rather than absolute concept. This new way of thinking did not overturn long-held understandings of puberty and its variable nature, particularly along social and moral lines, but provided a scientific basis for such claims. Statistics proved that puberty was a lengthy and variable process, meaning that few signs were inherently '(un)natural' or '(ab)normal'. Sexual forensics drew on the notion that certain *possible* explanations for bodily signs emerged at particular ages, and that these possibilities multiplied at puberty, but that few firm assumptions could be made about sexual maturity and capacity at any given age.

The ambiguities of sexual development opened up questions about the meaning of complainants' bodily and behavioural signs rather than providing clear-cut answers, but therein lay the value of sexual forensics. As *The Lancet* noted in 1905, in a published letter by the medical practitioner C. Bell Taylor of Nottingham, '[m]edicine is not an exact science and never will be and the man who comes into court to expound scientific truth to the laity is expounding something that does not exist'.[1] Ambiguity was part of medical knowledge and a knowledge practice. Medical witnesses tolerated ambiguity and perhaps even embraced it because, as Mark Jackson shows in his work on eighteenth-century infanticide trials, gaps in medical testimony were useful for judges and juries in a range of contexts.[2] Nancy Tuana has also shown in her work on 'the epistemology of ignorance' that ignorance 'is not a simple lack. It is often constructed, maintained, and disseminated and is linked to issues of cognitive authority, doubt, trust, silencing, and uncertainty'.[3] Although some scholars have claimed that 'experts' are under pressure to 'profess greater certainty than they really feel' in order to gain influence in court, in Victorian and Edwardian England judicial processes encouraged forms of scientific knowledge that left space for interpretation.[4]

Sexual forensics provided a way for the courts to open up discussions around the most problematic group of complainants: working-class pubescent girls. This 'problem' group came potentially to encompass a high proportion of complainants, particularly as physiology and anthropometric studies showed the wide age range at which puberty could occur. Medical emphasis on the ambiguities of physical development also allowed medical witnesses to highlight cases in which girls were mature in physical and/or behavioural, but not necessary legal, terms. The judicial process encouraged and shaped such medical testimony on sexual development, history and character. Defence lawyers often elicited testimony about a complainant's precocity or (un)chastity, while magistrates and grand juries prevented cases from reaching trial that did not fit with models of 'real rape'; a 'real' female victim was chaste and innocent, either childlike or modest in her outward asexuality. These stereotypes of victimhood continued to be prevalent, even after the age of consent changed to make most pubescent girls 'children' in the eyes of the law.

A pervasive focus on the sexual character of girls, even those under the age of sexual consent, was in part the consequence of general judicial conservatism and the consistency of courtroom 'scripts'. It was also the result of relatively slow attitudinal change in comparison with changes to sexual consent law. If anything, increases in the age of sexual consent actually fuelled concerns about blackmail and led to greater scrutiny of working-class girls' sexual histories. Medical research encouraged such concerns by highlighting the grey areas between childhood and adulthood. In 1885 the law on sexual consent defined girls below the age of 16 as lacking the capacity to consent to sexual activity, but medical research showed the extensive nature of precocity and the variable nature of sexual capacity. Physiology and medical jurisprudence texts identified gaps: between physical and mental maturity; between the growth of the sexual function and the ability to control it; and between 'biological' and 'social' puberty. Books of medical jurisprudence also increased the burden of proof on girls at puberty, both physical and behavioural, in response to growing fears about unfounded allegations. These shifts drew upon wider social concerns, in which there was both a spotlight on child protection and anxieties about false claims from older girls.

The value of sexual forensics lay in its support for growing middle-class concerns about age, gender and sexuality. These discussions did not fit neatly with new legal categories, but were equally a product of their time. Medical concerns about the ambiguities of sexual development and consent addressed broad social issues such as self-control, immorality and female modesty. They connected, for example, with social purity campaigns of the late-nineteenth century. Although social purity campaigners encouraged a rise in the age of sexual consent, they also repeatedly articulated concerns about working-class sexualities; in this framework pubescent girls *were* in danger, but the threat to their virtue came as much from their own emerging sexualities as from strangers on the street. The ostensible 'innocent/dangerous' and 'victim/threat' constructions of childhood, so often cited by historians, were thus not so much binaries as mutually dependent. Children also were not a homogeneous group and, for many, these two stereotypes related to distinct life stages. The particular concern about girls at sexual maturity was not limited to sexual forensics, but medical witnesses were able to present such concerns in a more scientific way by using developmental 'norms' to identify puberty and precocity.

In the late-nineteenth and early-twentieth centuries there was a growing medical recognition of the heterogeneous nature of childhood, with great changes between infancy and the long period of puberty. Medical witnesses drew upon a general social belief in the inherent innocence of infants and the very young, for whom conviction rates remained high throughout the period. Despite trends in some fields of medicine, particularly sexology, to engage with child sexuality such theories did not enter mainstream medicine until later in the twentieth century. Medical writers and witnesses situated any discovery of precocity against the middle-class 'norm' of innocence to claim that precocity was an atypical and undesirable state and to preserve childhood innocence. At the same time, they argued that precocity – both physical and behavioural – was more prevalent among the working classes because of the negative influence of heredity and environment.

Whether precocious or within the wide 'normal' age range, puberty was thought to be a time of significant biological and social disruption. It was lengthy, variable, multifaceted, unpredictable and problematised legal attempts to delineate clearly 'childhood' from 'adulthood'. As physical development was generally expected to precede psychological maturity, puberty also marked a period of potentially uninhibited sexuality. Concerns about uncontrolled sexuality at puberty related to girls and boys alike, but fed into sexual forensics in different ways according to gender. Female bodies were a particular focus of sexual assault trials, of medical literature on sexual development and of social concerns about girls' sexual character. Medical testimony also paid attention to the emotional 'performance' of children and female complainants. As Garthine Walker notes, there was a growing mistrust of complainants with the emergence of the adversarial system.[5] The adversarial system, and forensic medicine within it, placed an increasing burden upon female complainants to adhere to age-based and gendered stereotypes of 'victimhood'.

Overall, sexual forensics was part of what Shani D'Cruze and Louise Jackson describe as the 'staged event' of the criminal trial.[6] Courtroom 'scripts' had particular narrative conventions that at first glance seem surprisingly consistent across time and place. Models of 'real rape' changed little between 1850 and 1914, as did the medical evidence that related to such crimes. The first case tried in this period was a Middlesex trial from 1850, for an alleged offence on a nine-year-old girl, in which the medical witness stated:

> I am a Surgeon and reside at No. 42 Downham Row, Islington. About ten days or a fortnight ago the child Jane Dorrington was brought to me by it's [sic] father, and I examined the private parts of the child and I found the genitals very much swelled and a profuse discharge. There was every appearance of violence of some mechanical nature having been used to the child.[7]

In a much later case of alleged assault on a 'little girl', from 1913, the medical witness stated that:

> I am M.B. and a reg^d medical practitioner practising at Banwell. I was called to see Ruby Glover shortly before 10 a.m. on 25th August. She told me her story and, with her father's consent, I examined her. There were no marks of bruising on her body and no evidence that she had been violated or any attempt. There were no stains or bloodmarks on the clothes. The drawers were not torn or damaged.[8]

There was no significant difference between these two statements. Both witnesses were local general practitioners, without forensic expertise. They focused on physical signs and gave relatively short and descriptive testimony, based on naked-eye diagnosis rather than chemical or microscopic analysis. The final case with surviving records from the period related to a girl aged 16, also from Somerset, in which the medical witness testified that:

> I am a duly registered medical Practitioner at Highbridge – M.B. Dublin – On Thursday 19th March instant I saw Beatrice Webb at Mr Puddy's house Brent Knoll. With her consent and in the presence of her mother I examined her. I found nothing abnormal. I examined her private parts. There was no evidence of any injury. No trace of any penetration in the vagina and no bruising about the thighs. She made a statement to me.[9]

The only significant shift between first and last testimony was that the final medical witness referred explicitly to the 'abnormal' body. The first testimony focused on the question of 'mechanical' violence, a term implicitly in opposition to the 'natural' body, but later testimony focused more on the 'normal' body. This final medical testimony was representative of a subtle shift towards the greater use of 'normality' as an explicit framing device over the late-nineteenth and early-twentieth centuries. Otherwise, the language and subject matter of sexual forensics in court was relatively stable.

This relative stability seems remarkable over more than half a century, especially as this period – albeit towards the end – included growing recognition of sexology and psychoanalysis, challenges to patriarchy and dramatic changes to sexual consent law. As so many concerns about sexual immorality were articulated in relation to urban areas, with London as a focus of anxieties about dirt and disorder, it also seems surprising that medical witnesses, judges and juries demonstrated as much concern about the working-class girls in the provinces and rural areas as they did about working-class females based in big cities. However, the lack of significant change in courtroom 'scripts' does not mean that they did not respond at all to wider social trends. Shifts within the practice of sexual forensics, and in the courts more generally, were more low-key and gradual than in the outside world. Medical testimony shifted subtly in relation to bodily 'norms', for example, and some courts moved slowly towards a less gendered model of consent in response to social challenges to the 'double standard'. Other concerns, such as those relating to precocity, were period-specific and represented contemporary concerns about uncontrolled working-class sexualities. New medical theories about physiology – from sexual maturity to the nerves – were integrated into courtroom 'scripts' almost invisibly, in a way that reinforced rather than challenged the status quo. This trend was not limited to England. Dan Healey, one of the few other scholars to have researched sexual forensics in history, found in his work on Bolshevik Russia that a so-called 'sexual revolution' had no immediate impact on the patriarchal court system.[10] Although Healey's work focuses on a later period, his findings echo those of this book about the conservative and somewhat self-reinforcing nature of sexual forensics in the courts throughout history. New medical theories and social concerns entered the judicial sphere in a piecemeal fashion; there was also no 'sexual revolution' in the courtroom.

One of the most important, but often overlooked, aspects of these 'scripts' relates to the age of complainants and prisoners. There has often been an implicit scholarly assumption that gender is the primary axis upon which sexual crime turns, and that there is one model of 'real rape' and another of 'child sexual abuse'. Conversely, sexual forensics shows that there were multiple models of 'real rape' and 'victimhood' that – although they shared key characteristics about class and respectability – turned as much on age as on gender. Throughout the Victorian and Edwardian periods the legal delineation between 'carnal knowledge' of children and 'rape' of adults did not operate so clearly in practice. Children were not a homogeneous group and the most 'complex and controversial' trials related to girls at the intersection between 'childhood' and 'adulthood'. At the lengthy and variable life stage of puberty, complainants and prisoners alike embodied contemporary anxieties about uncontrolled working-class sexualities. Looking at sexual crime through the lens of age indicates that, in the words of Sally Shuttleworth, 'it [is] now time to add age … to the triumvirate of class, gender and race'.[11]

Moving beyond a framework that separates 'children' from 'adults' opens up new questions and provides valuable insights not only into the history of sexual forensics, but also into histories of puberty, sexuality and society.

Notes

1. C. Bell Taylor, 'The Medical Man as Expert Witness', *The Lancet*, 1 April 1905, 887–88, p. 887.
2. Mark Jackson, *New-Born Child Murder: Women, Illegitimacy and the Courts in Eighteenth-Century England* (Manchester: Manchester University Press, 1996), p. 15.
3. Nancy Tuana, 'Coming to Understand: Orgasm and the Epistemology of Ignorance', *Hypatia* 19 (2004), 194–232, p. 194. I am grateful to Willemijn Ruberg for drawing my attention to this subject of 'epistemologies of ignorance' and for sending me an unpublished version of her article to do so.
4. Anthony Good, *Anthropology and Expertise in the Asylum Courts* (Abingdon; New York: Routledge, 2007), p. 45.
5. Garthine Walker, 'Rape, Acquittal and Culpability in Popular Crime Reports in England, c.1670–c.1750', *Past and Present* 220 (2013), 115–42.
6. Shani D'Cruze and Louise A. Jackson, *Women, Crime and Justice in England since 1660* (Basingstoke; New York: Palgrave Macmillan, 2009), p. 12.
7. London, London Metropolitan Archives (LMA), Pre-Trial Statements, George Warren tried at the Middlesex Sessions on 29 August 1850 for indecent assault, MJ/SP/E/1850/016.
8. Taunton, Somerset Heritage Centre (SHC), Pre-Trial Statements, William Murray tried at the Somerset Quarter Sessions on 16 October 1913 for indecent assault, Q/SR/853.
9. Taunton, SHC, Pre-Trial Statements, William Adams and Stanley Lewis Crandon tried at the Somerset Quarter Sessions on 8 April 1914 for indecent assault, Q/SR/855.
10. Dan Healey, *Bolshevik Sexual Forensics: Diagnosing Disorder in the Clinic and Courtroom, 1917–1939* (DeKalb: Northern Illinois University Press, 2009).
11. Sally Shuttleworth, 'Victorian Childhood', *Journal of Victorian Culture* 9 (2004), 107–13, p. 107.

Selected Bibliography

Manuscript primary sources

Exeter, Devon Record Office, Pre-Trial Statements 1850–1914, QS/B/1850–1914.
Gloucester, Gloucestershire Archives, Pre-Trial Statements 1850–1914, Q/SD/2/1850–1914.
London, London Metropolitan Archives, Pre-Trial Statements, 1906–09, MXS/B/01/003.
———. Pre-Trial Statements, 1903–1906, MXS/B/01/002.
———. Pre-Trial Statements, 1890–1903, MXS/B/01/001.
———. Pre-Trial Statements, 1850–1889, MJ/SP/E/1850–1889.
Taunton, Somerset Heritage Centre, Pre-Trial Statements, 1850–1914, Q/SR/575–857.

Printed primary sources

Acton, William, *The Functions and Disorders of the Reproductive Organs in Youth, in Adult Age, and in Advanced Life, Considered in their Physiological, Social and Psychological Relations* (London: John Churchill, 1857).
Blackwell, Elizabeth, *The Human Element in Sex: A Medical Inquiry into the Relation of Sexual Physiology to Christian Morality* (London: J. & A. Churchill, 1894).
Bristol (Daily) Mercury, 1850–1909.
British Medical Journal (BMJ), 1857–1914.
Carpenter, William B., *Principles of Human Physiology: With their Chief Applications to Psychology, Pathology, Therapeutics, Hygiene & Forensic Medicine*, 5th edn (London: John Churchill, 1855 [1842]).
Casper, Johann Ludwig, *Handbook of Forensic Medicine*, trans. from 3rd edn by G. W. Balfour, 4 vols (London: New Sydenham Society, 1861–65).
Cox Criminal Law Cases, 1850–1914.
Criminal Appeal Reports, 1909–1914.
Drysdale, George R., *The Elements of Social Science; or Physical, Sexual and Natural Religion. An Exposition of the True Cause and Only Cure of the Three Primary Social Evils: Poverty, Prostitution, and Celibacy*, 27th edn (London: E. Truelove, 1889 [1854]).
Dyer, Alfred S., *Facts for Men on Moral Purity and Health: Being Plain Words to Young Men upon an Avoided Subject with Safeguards against Immorality, & Facts that Men Ought to Know* (London: Dyer Brothers, 1884).
Edinburgh Medical (and Surgical) Journal, 1844–1914.
Ellis, Havelock, *Studies in the Psychology of Sex: Analysis of the Sexual Impulse, Love and Pain, the Sexual Impulse in Women*, vol. 3 (Philadelphia: F. A. Davis Company, 1904 [1903]).
Exeter, Devon Record Office, Calendars of Prisoners, 1850–1914, Q/S/32/192–5853.

Gloucester, Gloucestershire Archives, Calendars of Prisoners, 1850–1914, Q/SGc/ 1850–1914.
Guy, William A., and David Ferrier, *Principles of Forensic Medicine*, 5th edn (London: Henry Renshaw, 1881 [1844]).
Hall, G. Stanley, *Adolescence: Its Psychology and its Relations to Physiology, Anthropology, Sociology, Sex, Crime, Religion and Education*, 2 vols (London; New York: Appleton, 1904).
HC Deb. (3rd–5th series), 1850–1914.
HL Deb. (3rd–5th series), 1850–1914.
Illustrated Police News, 1864–1914.
Krafft-Ebing, R. v., *Psychopathia Sexualis*, trans. C. G. Chaddock, 10th edn (London: Rebman, 1899 [1886]).
Lancet, 1850–1914.
Leeds Mercury, 1885.
Lloyd's Weekly Newspaper, 1850–1914.
London, London Metropolitan Archives, Calendars of Prisoners, 1906–1915, MXS/B/03/002.
———. Calendars of Prisoners, 1889–1905 MXS/B/03/001.
———. Calendars of Prisoners, 1855–1892, MJ/CP/B/001–043.
———. Calendars of Prisoners, 1850–1853, MJ/CP/A/031–069.
Law Reports, Crown Cases Reserved, 1865–1875.
Mead, Frederick, *The Criminal Law Amendment Act, 1885, with Introduction, Notes and Index* (London: Shaw & Sons, 1885).
Morning Chronicle, 1850–1862.
Morning Post, 1850–1914.
Ogston, Francis, *Lectures on Medical Jurisprudence* (London: J. & A. Churchill, 1878).
Pall Mall Gazette, 1885.
Provincial Medical & Surgical Journal, 1840–1852.
Reynolds's Weekly Newspaper, 1850–1914.
Tait, Lawson, 'An Analysis of the Evidence in Seventy Consecutive Cases of Charges made under the New Criminal Law Amendment Act', *Provincial Medical Journal*, 1 May 1894, 226–33.
Taunton, Somerset Heritage Centre, Calendars of Prisoners, 1882–1914, A/CJA/1/1-16.
———. Calendars of Prisoners, 1850–1882, Q/SCS/1–330.
Taylor, Alfred Swaine (*Manual of*) *Medical Jurisprudence*, 1st–12th edns (London: J & A Churchill, 1844–1891).
———. *The Principles and Practice of Medical Jurisprudence*, 1st–6th edns, 2 vols (London: J. & A. Churchill, 1965–1910).
The Standard, 1850–1914.
The Times, 1850–1914.

Published secondary sources[1]

Arnot, Margaret L., and Cornelie Usborne (eds), *Gender and Crime in Modern Europe* (London: UCL Press, 1999).
Bland, Lucy, *Banishing the Beast: English Feminism and Sexual Morality 1885–1914* (London: Penguin, 1995).

Bourke, Joanna, *Rape: A History from 1860 to the Present Day* (London: Virago, 2007).
Brownmiller, Susan, *Against Our Will: Men, Women and Rape* (Harmondsworth: Penguin, 1977 [1975]).
Clark, Michael, and Catherine Crawford (eds), *Legal Medicine in History* (Cambridge; New York: Cambridge University Press, 1994).
Conley, Carolyn A., *The Unwritten Law: Criminal Justice in Victorian Kent* (Oxford: Oxford University Press, 1991).
D'Cruze, Shani, and Louise A. Jackson, *Women, Crime and Justice in England since 1660* (Basingstoke; New York: Palgrave Macmillan, 2009).
Davis, Joseph E., *Accounts of Innocence: Sexual Abuse, Trauma, and the Self* (Chicago; London: University of Chicago Press, 2005).
Digby, Anne, *The Evolution of British General Practice, 1850–1948* (Oxford: Oxford University Press, 1999).
Douglas, Mary, *Purity and Danger: An Analysis of the Concepts of Pollution and Taboo* (London: Routledge and Kegan Paul, 1978).
Eigen, Joel Peter, *Unconscious Crime: Mental Absence and Criminal Responsibility in Victorian London* (Baltimore; London: Johns Hopkins University Press, 2003).
Ernst, Waltraud (ed.), *Histories of the Normal and the Abnormal: Social and Cultural Histories of Norms and Normality* (Abingdon; New York: Routledge, 2006).
Fisher, Kate, and Sarah Toulalan (eds), *Bodies, Sex and Desire from the Renaissance to the Present* (Basingstoke: Palgrave MacMillan, 2011).
Foucault, Michel, *The History of Sexuality, Vol. 1. An Introduction*, trans. Richard Hurley (New York: Pantheon, 1978).
Garton, Stephen, *Histories of Sexuality* (London: Equinox, 2004).
Hacking, Ian, *The Taming of Chance* (Cambridge: Cambridge University Press, 1990).
Healey, Dan, *Bolshevik Sexual Forensics: Diagnosing Disorder in the Clinic and Courtroom, 1917–1939* (DeKalb: Northern Illinois University Press, 2009).
Jackson, Louise A., *Child Sexual Abuse in Victorian England* (London: Routledge, 2000).
Mason, Michael, *The Making of Victorian Sexuality* (Oxford; New York: Oxford University Press, 1994).
Micale, Mark S., and Paul Lerner (eds), *Traumatic Pasts: History, Psychiatry, and Trauma in the Modern Age, 1870–1930* (Cambridge: Cambridge University Press, 2001).
Porter, Roy, and Sylvana Tomaselli (eds), *Rape: an Historical and Cultural Enquiry* (Oxford: Basil Blackwell, 1986).
Robertson, Stephen, *Crimes against Children: Sexual Violence and Legal Culture in New York City, 1880–1960* (Chapel Hill; London: University of North Carolina Press, 2005).
Shuttleworth, Sally, *The Mind of the Child: Child Development in Literature, Science and Medicine, 1840–1900* (Oxford: Oxford University Press, 2010).
Walkowitz, Judith R., *City of Dreadful Delight: Narratives of Sexual Danger in Late Victorian London* (Chicago: University of Chicago Press, 1992).
Watson, Katherine D., *Forensic Medicine in Western Society: A History* (London; New York: Routledge, 2010).
Weeks, Jeffrey, *Sex, Politics and Society: The Regulation of Sexuality since 1800* (London: Longman, 1981).

Note

1. This short list of 'selected' secondary reading gives only monographs or edited collections that relate broadly to the subject matter of the book as a whole, for the purposes of further reading. There are many excellent, more specific works or book chapters and journal articles that are also referenced in footnotes and in this book's main text.

Index

Adolescence (*see also* puberty and Hall, G. Stanley) 9–10, 145
Age of consent 2, 6–7, 9, 10–16, 28, 31, 55, 68, 82, 86, 91, 93, 109, 113–14, 118, 127, 159, 160, 163–4, 191–2
  1861 Offences Against the Person Act 13, 62
  1875 Offences Against the Person Act 13, 107, 127
  1880 Criminal Law Amendment Act 13, 109, 120, 159
  1885 Criminal Law Amendment Act 13–15, 50, 91, 107, 110–11, 115, 117, 163–6, 179, 192
  international 14–15
  male 14, 159–60

Blackmail: *see* false claims
Blood (*see also* menstruation) 22, 24, 26, 32, 36, 42–4, 49–52, 79–80, 89, 98, 138, 190, 194
Bruising 44, 47, 79, 88–9, 113–16, 194

Carnal knowledge – law on: *see* 1861 Offences Against the Person Act, 1875 Offences Against the Person Act, and 1885 Criminal Law Amendment Act
Case analyses
  consent 124–9
  emotions 149–53
  injury 65–70
  offenders 183–5
  unchastity 95–100
Case law 106, 108–11, 160
Child protection movements 7, 9–11, 16, 55
Child sexual abuse 11–12, 134, 158, 182, 195
Crying 135–8

Dilation, genital 24, 32, 36, 43, 44, 47, 88

Expertise 2, 22–4, 42–3, 76, 93, 118, 137–8, 140–1, 191, 194

False claims 14, 33, 88, 91, 93, 105, 115, 147–8, 153, 163–5
Fear 107–8, 133, 140, 143
Femininity 8, 43, 86, 90–1, 94, 108, 132, 141
Freud, Sigmund 81, 148

Gonorrhoea: *see* venereal disease

Hall, G. Stanley 9–10, 134, 145
Hymen 57, 66, 76–9, 82–3, 85, 88–90, 92
Hysteria 132–3, 141, 143–4, 146–53

Imbecility 110, 117, 124–9, 169, 172–4, 178
Impotence 159–61, 176
Indecent assault – law on: *see* 1880 Criminal Law Amendment Act
Incest 13–14, 77, 108, 110, 121–22, 170, 180–2
Inflammation 43, 47–9, 60, 66–9, 184
Insanity 152, 168–9, 172–3, 178
Insensibility 106, 116, 120–1

Laceration 43–4, 46–7, 60, 66–7, 82, 84, 89, 92, 160

Masculinity 11, 84, 136, 158, 162, 167–70, 173
Masturbation 79–85, 89, 92, 145
Medical jurisprudence 24–5, 30–3, 42–4, 53–5, 76, 83, 85, 107, 109, 111–13, 116, 133, 175–6, 179, 181, 192

Menstruation (*see also* blood) 25–8, 31–2, 34, 36, 49–52, 80, 99, 141–2, 145, 152–3

Nervous conditions 34, 138–48, 152–3
Newspapers 3, 12, 58–63, 78, 86–92, 116, 118–22, 132, 145, 147–8, 153, 164, 167–8, 171–3, 176, 181
Normality 8, 10, 15, 23–36, 43, 46–52, 55, 63, 68–9, 78, 81, 88, 105, 112, 114, 128, 137, 142, 148, 153, 161, 163, 171–8, 180–1, 184, 190, 192–5

Old Bailey 4, 7, 22–3, 32, 35–36, 128

Paedophilia 12, 158, 171, 177, 179, 182
Physiology 9, 16, 23–5, 29–31, 42, 46, 52, 56–7, 83, 86, 110–12, 134, 138, 140, 145, 148, 152–3, 161, 176, 181, 190–2, 195
Precocity 7, 15–16, 28–9, 31, 33, 44, 46, 48, 50–1, 78, 80–1, 83, 86, 88, 95, 112–13, 129, 163–4, 174, 191–3, 195
Pregnancy 31, 43, 50–1, 112–13, 145, 150–2
Psychiatry 141, 143, 145, 169, 173, 178
Psychology 9–10, 25–6, 83–4, 105–7, 118, 134–5, 144–5, 148, 194

Puberty (*see also* adolescence) 6, 9, 11–13, 15, 24–8, 30–2, 42–3, 46–7, 49–52, 54, 56–7, 63–4, 76, 82–6, 88–95, 106, 109–14, 136, 141–2, 145, 160, 162–7, 169–70, 174–5, 179, 190–3, 195–6

Rape – law on 44, 57, 85, 105–7, 109–10, 118, 159–60
Rape myths 3, 6, 15–17, 32, 35, 87, 100, 106, 114, 123, 153
Respectability 7, 58, 70, 83, 85, 87, 90–2, 99, 107–8, 112, 119, 132, 142, 164, 167, 181, 195

Sentencing 16, 22, 52, 62–3, 92–3, 121–2, 143, 159, 167–9, 177, 185
Sexology 26, 94, 158, 170–1, 181–2, 193–4
Sodomy 5, 14, 88, 159, 171
Sperm 26, 66, 68, 84, 144–5
Stereotypes 6, 15–16, 34–5, 76, 86–7, 90–1, 93–4, 99, 106, 122, 133, 137, 140–1, 177–9, 184, 191–3
Stead, W. T. 12–13, 55
Syphilis: *see* venereal disease

Trauma 133–4, 144–5

Venereal Disease 43, 52–6, 63, 65–70
Verdicts 6, 48, 59–62, 64, 91–3, 110, 118–22, 146–8
Virginity: *see* hymen